Critical Human Geography

'Critical Human Geography' is an international series which provides a critical examination and extension of the concepts and
consequences of work in human geography and the allied social
sciences and humanities. The volumes are written by scholars
currently engaged in substantive research, so that wherever possible
the discussions are empirically grounded as well as theoretically
informed. Existing studies and the traditions from which they derive
are carefully described and located in their historically specific
context, but the series at the same time introduces and explores new
ideas and insights from the human sciences as a whole. The series is
thus not intended as a collection of synthetic reviews but rather as a
cluster of considered arguments, which are accessible enough to
engage geographers at all levels of the subject in its development. It
therefore reflects the continuing methodological and philosophical
diversity of the subject, and its books are united only by their
common commitment to the prosecution of a genuinely human
geography.

Department of Geography   MARK BILLINGE
University of Cambridge   DEREK GREGORY
England   RONALD MARTIN

Critical Human Geography

Conceptions of Space in Social Thought: a Geographic Perspective
Robert David Sack

FORTHCOMING TITLES

A Cultural Geography of Industrialisation in Britain
Mark Billinge

Geography as Spatial Science: Recollections of a Revolution
Mark Billinge, Derek Gregory and Ronald Martin (editors)

Geography and the State: an Introduction to Political Geography
Ron Johnston

Regions and the Philosophy of the Human Sciences
Nicholas Entrikin

Between Feudalism and Capitalism
Robert Dodgshon

Capital and the Space Economy
Michael Dunford and Diane Perrons

Conceptions of Space in Social Thought

A Geographic Perspective

Robert David Sack

M

First published 1980 by
THE MACMILLAN PRESS LTD
London and Basingstoke
Associated companies in Delhi Dublin
Hong Kong Johannesburg Lagos Melbourne
New York Singapore and Tokyo

Printed in Hong Kong

British Library Cataloguing in Publication Data

Sack, Robert David
    Conceptions of space in social thought.
    – (Critical human geography).
    1. Space perception
    I. Title   II. Series
    153.7'52   BF469

    ISBN 0-333-28683-9
    ISBN 0-333-28684-7 Pbk

To Karen

Contents

Part III Unsophisticated–fused Patterns

Part IV Unsophisticated and Sophisticated Patterns

Part V Conclusion

List of Illustrations

Acknowledgements

I am grateful for the receipt of a John Simon Guggenheim Memorial Foundation Fellowship and a Wisconsin Alumni Research Foundation Fellowship from the Graduate School of the University of Wisconsin, Madison. The support from these foundations allowed me to devote the 1975–6 academic year to formulating and researching the problems discussed in this book. I want to thank the following individuals for reading earlier drafts of the manuscript and for offering excellent criticisms and most needed encouragement: Martin Cadwallader, Daniel Doeppers, Muriel Dury, Nick Entrikin, Fred Lukermann, Charles Mahaffey, Karen Sack, Yi-Fu Tuan and David Ward. I want to thank the following individuals and organisations for their copyright permission and co-operation: the Meadows Museum, Southern Methodist University, Dallas, Texas, for Portrait of Philip IV by Velázquez; Frau G. Niedieck, Zurich, for the Yantra mandala; Trustees of the Chatsworth Settlement and the Courtauld Institute of Art for 'Et in Arcadia Ego' by Poussin; Gustav Fisher Verlag, Jena, for W. Christaller's 'Central Places in Southern Germany', in *Die Zentralen orte in Suddeutsthland* (1933); the National Gallery, London, for 'Portrait of Philip IV' by Velázquez; Professor P. Wheatley for Giwang-d̂ˌiĕng from *Pivot of the Four Quarters*; the Louvre, Paris, for 'Les Bergers d'Arcadie' by Poussin; the Library of Congress for the 10th century T–O map from Santarem's *Atlas Compose de Mappemondes*.

Part I
Introduction

1
Space and Modes of Thought

Geography and space

Since antiquity geographers have explored and analysed the earth's surface from two related perspectives: that of the spatial differentiation and association of phenomena with an emphasis on the meaning of space, spatial relations and place; and that of the relationship between man and his physical environment. The two are closely related because the meanings of space and place depend on the interrelationships among physical and human activities located in space, and man's relationships to the environment occur in the context of space and place. The two emphases come together in the idea of landscape and human impact on the land.

This book is about the spatial perspective primarily. We will analyse the nature of terrestrial or geographic space, bearing two major issues in mind. The first is that geographic space is seen and evaluated in different ways at different times and in different cultures. That is to say, there are different meanings or conceptions of space. Our first and most important task is to characterise and analyse these conceptions and their interrelationships. We will do so in terms of a realist framework which can be used to explore the human as well as the physical elements on the earth's surface.[1] The second issue stems from a critical evaluation of the first. It concerns the adequacy of this perspective as a framework for embracing and elucidating these different meanings of space. Analysing the extent to which a realist science embraces them, and pointing to the problems it and narrower conceptions of science have encountered, lays a foundation for a critique of science from a spatial perspective. This is needed now especially, if we consider the recent awareness of the multiple meanings of space and place, the doubts about the adequacy of (often narrowly conceived) scientific method in human geography and in social science in general, and the increasing awareness by geographers and social scientists of a number of alternative approaches to the study of man as expressed in human geography's recent interest in structuralism, phenomenology, idealism, Marxism, and many valued logics.[2] Our intent here is not to review these very important positions and reconcile them but rather to consider the capacity of a broadly conceived realist scientific framework to incorporate the multiple conceptions of space raised by our awareness of different viewpoints and philosophical approaches to human behaviour.

Meanings of space

Space is an essential framework of all modes of thought. From physics to aesthetics, from myth and magic to common everyday life, space, in conjunction with time, provides a fundamental ordering system interlacing every facet of thought. We are constantly reminded of the function of space when we use such expressions in ordinary language as 'Everything has its *place*', or 'To which one are you referring, this one *here*, not that one *there*?' The *here*, the *there*, the *place* refer to part of a spatial framework for knowledge about the world. In short, things occur or exist in relation to *space* and time. Even the definitions of these words are closely connected. A thing is defined as an event, fact or occurrence. To occur is to happen, to befall or to take *place*. To exist is to have being in relation to *space* and time. This entanglement of space and thing and space's infusion into every realm of thought makes it a particularly important, but difficult, concept to isolate and analyse. The problem of analysing space is compounded because whatever may be said of space at the level of theoretical physics and philosophy, at the terrestrial level, geographic space is not empty. It is filled with matter and energy, or substance. The fact that people discuss this space, describe it, and analyse it, means that they are conceptually – not actually – isolating and separating space from substance. When we refer to space and its properties in this book we are referring therefore to a system which is conceptually, not actually, separable from facts and their relationships.

   Different conceptions of space arise because this conceptual relation and separation can occur at different levels of abstraction and from different viewpoints and modes of thought. Within each level and mode it can occur differently and have different import or significance in terms of its relationship to feelings and the facts from which it is abstracted. Very roughly we can think of the process of conceptual separation along a continuum beginning with perception, then on to description, and ending with analysis and evaluation. These categories of course are not independent. What we perceive depends in part on how we evaluate, analyse and describe, and certainly our descriptions, evaluations and analyses depend on our perceptions. Perception itself involves abstraction in the sense that we selectively receive, recognise and respond to only some of the stimuli in the world. But abstraction most clearly occurs in our descriptions, evaluations and analyses, and these to a great degree determine the conceptions of space. While conceptions of space are clearly about abstractions, they also involve perceptions of spatial relations and how these are described.

Space often changes its meanings because we perceive and describe the spatial relations among things differently in different situations. Distances, patterns and shapes literally take on different appearances and can be described differently from different viewpoints. From one point of view a circle may appear as an ellipse; from another it may not be discernible as a closed figure, if discernible at all. Differences in the way space and its properties appear to us and in our descriptions of such appearances may be due to differences in our attentions to detail, to our cultural environments, to our access to technology and education, and to our stages in the life cycle. The appearance of space is also altered by our being blind, or deaf, or otherwise biologically handicapped, or by our having immature sensory systems as in the very young child whose perceptions of spatial relations are less complete than the adult's.

Geography, through cartography, has already done much to coordinate some of the differences among perceptions and de-scriptions of space that result from different technological levels, ages, personal orientations and degrees of abstractions. Cartographic projections, scales and controlled distortions can portray different views of spatial relations. Because these projections, scales and distortions are systematically interrelated, the map can coordinate personal views and perspectives with a standard yet flexible de-scription of space. This description provides a frame of reference abstracted from the material world and with regard to which the material objects and their relationships can be identified and described. The map allows geography to perform its most basic function of standardising perspectives of space and providing objective descriptions of the spatial properties of things, just as chronology allows history to provide its essential and elementary function of objectively recording the temporal order of things.

Readily accessible knowledge of precise location with regard to physical space and time is fundamental for the unequivocal identification and individuation of things.[3] Without it, we would find it extremely difficult to have clear, factual discourse. We use a space–time system to say to which of several apparently identical things one is referring by locating a thing with reference to space and time. If the space–time system were a limited and personal one, that discourse would be limited to people who were in the same place and who could see the same things. Problems of identification would be solved by pointing and saying, 'I mean that one over there'. If the potential conversants were to be apart there would be no way to bring them together without having them, or a third party, possess an objective perspective of spatial relations, the kind of perspective which a map provides.

Providing a description of an abstract spatial system and then identifying things by their location within it is by no means all there is to geography (just as description in a temporal framework is not all there is to history) or to the role of space. Space enters into the analyses and evaluations of things in the way it is thought to affect things. The different views of the effects of space are the major sources of different conceptions of space. Fundamental differences in the evaluation of even the same spatial configuration can arise when the configuration is examined from one or another of such modes as the scientific, the social-scientific, the artistic, the mythical–magical, the child's, the practical and the societal. From a scientific perspective, the circle is the most compact two-dimensional figure. This explains why many social systems, such as settlements, which need to minimise accessibility, often tend to be circular. But this same shape, although seen as a circle in other modes, can be evaluated differently. It can be an object of aesthetic contemplation, a symbol of perfection, or a form possessing magical powers. And the scientific concept of 'compactness' may not readily explain these other interpretations. Such changes in meaning occur for other shapes and spatial properties as well. Individually and collectively, the evaluation of space in such modes as the mythical–magical, the scientific, the social-scientific, the artistic, the societal, the child's and the practical have moulded our behaviour in space, our built environments, our alterations of the earth's surface and our responses to the landscape. These clearly are geographic effects of the conceptions of space.

Since different perceptions and especially descriptions and evaluations or analyses result in different conceptions of space, we may, through our explorations, easily come to conflicting terminologies and ambiguity. To help avoid this, it is important to find a perspective which leads to a description and an analysis or evaluation of space that is broad, clear, and well-known, and that can be used as a standard for the other meanings. Given the enormous influence of science in our culture, and geography's position as a physical and social science, we have naturally turned to the physical sciences for such a view of space.

As a brief introduction to this complex view we may say that according to it, space means the space of the physical world, and geometry is the language in which it is described. Description imparts meaning, and geometric description has helped determine one of the two important scientific meanings of physical space. Although the appropriate geometry for those activities at the extreme ends of the physical universe, the sub-atomic particles and the astronomical are still debated, science is agreed that for the intermediate scale which we usually refer to as geographic and in which are located the social and physical events of the earth's surface that constitute the immediately perceived and humanly created environment, the appropriate description of physical space is the familiar abstract three-dimensional Euclidean geometry in which a straight line may be drawn from any point to any other point, in which parallel lines do not meet at infinity, in which a circle may be described with any centre and any radius, and so on. This is (approximately) the description of physical and especially geographic space the physical scientist bequeaths to the geographer and to the social scientist. It is the framework in which their facts and events are supposed to be located.[4]

The other important scientific meaning of space stems from its evaluation, which in science is about the conceptual connection or recombination of the space with time and things or substances in laws and theories. The evaluation of space in science will be referred to as the *significance* of space. In the long run the scientific significance of space is related to the description of space, but for our purposes the two can be discussed separately.[5] Scientific significance, as the functional relationships among variables, is normally specified in science by the way in which the variables enter into relationships in laws and theories. The significance of space, its conceptual recombination with substance and time, therefore is about its inclusion in the laws of science. Although there are many kinds of laws, those about terrestrial events are expected to conform to general presuppositions about the role of space and time in causality; namely that an effect cannot precede a cause and that actions occur through causal chains in space.

The influence of science is so pervasive in the modern industrial west that a modified form of these presuppositions has uncritically and in part unconsciously seeped into the social sciences and even into our everyday practical expectations about the causes of things. The uncritical acceptance of these presuppositions has meant that they and their implications have not been well understood in social science. Clarification of the conception in social science though is no easy task. For one, it would require laws and theories of human action to make the significance of space for human behaviour more than hypothetical. But there are as yet few if any laws in social science that are comparable in scope and rigour to those of the physical sciences. For another, there is a lack of consensus about what constitutes explanation in social science. Although the social sciences have been greatly influenced by the methods of physical science, this influence has been received with mixed emotions and effects. Despite the assumption of their title 'scientists', social scientists are not of one mind about appropriate methods for studying human behaviour and about what would constitute laws, generalisations and explanations in their disciplines or even if a science of human behaviour would produce explanations. These methodological conflicts and vagaries have made it difficult to develop guidelines for the incorporation of physical space into social science. Moreover, other modes of thought such as art, myth and magic, and the child's view, give space a very different meaning. This fact alone would affect the social science meaning of space, for one of the objectives of social science has been the study of the way people 'see' the world. If people see, and/or evaluate things and space differently and 'non-scientifically', then social science must somehow represent and capture these meanings. Nevertheless it is possible to find the rudiments of a general social science view of space from what consensus does exist about the nature of social science and from an analysis of our everyday practical expectations about cause and effect. The social science and the practical views will be introduced together in this next section because the social sciences are not, yet, far removed from a practical commonsense view. A fuller articulation of these modes and a discussion of their differences will be the subject of subsequent chapters.

Space in practical and social science modes

Action by contact

The view of space in science that most affects the practical and social-scientific orientations in Western society is a conflation of the principles of action by contact and the conservation of energy. We will discuss these principles in more detail in Chapter 2, but we will begin here by pointing out that this conflation is the primary way in which space is conceptually, albeit mostly unconsciously and uncritically, recombined to substance and time in the social sciences and in the practical view. It is the way in which space and its properties can be seen to affect things. An important part of this conflation can be expressed in the following schema: if x influences y, x must either be in contact with y directly, by being physically in touch with it, or, if x and y are physically apart, the influence of x on y must travel through intervening substances (i.e., $a, b, c, \ldots n$) which form a medium, a path, or network of substances linking x to y. If such routes do not exist, we cannot, from this view, claim that x affected y. If such claims are insisted upon, they may indicate another mode of viewing space. Turning this around, properties of space such as location and distance by themselves have no effect on substances. Rather it is substances in space with spatial properties which affect other substances in space with spatial properties. While this conception of causality is a conflation of the principles of action by contact and the conservation of energy, for the sake of brevity, we shall refer to it simply as action by contact.

The way in which this concept influences our practical and social-scientific views of cause and effect is seen in the ease with which we accept a principle such as the theory of infectious diseases. The part of the theory of interest to us contends that infectious diseases are caused by living organisms and that these organisms are transmitted through specific media. The agent causing cholera, for instance, is the bacterium *vibrio cholerae*. It enters the body usually through the mouth when food or water containing the organism is ingested. Once inside the body it infects the mucous membranes lining the small intestines causing diarrhoea and vomiting. These evacuations contain the bacteria and without proper sanitary procedures may return through sewage to infect the water supply, or through insanitary food handling and preparation to infect the food supply, and further spread the disease among the population. There may be many permutations of substances or media in this chain, but the organisms causing the disease are known and in order to cause infection they must be carried by substances in which they can survive.

Both the plausibility and the discovery of this theory, which occurred as recently as the second half of the nineteenth century, are due in part to our acceptance of the scientific view of cause and effect. Before the causal agent was isolated, the supposition that it was an infectious disease, coupled with the concept of action by contact, helped to identify water as one of the media which transmits cholera. In the mid-nineteenth century the physician and scientist John Snow noticed that enormous outbreaks of cholera occurred within a small radius of one of the public water pumps in London while there were far fewer around the others. Furthermore, he found that those not living near the pump, but who had drunk from the water, either at nearby pubs or on visits to friends and relatives, came down with the disease in greater proportion than those who had no contact with the pump. Such associations with a particular substance (water) at a particular place led Snow to conclude that cholera was transmitted by water and that the water could be tainted by sewage.[6]

But not everyone accepted or even sought such causal links. Some believed cholera to be a product of the local environment, 'bad air', or of unnatural or supernatural forces. The farther back in time one goes, the more dominant may have been the supernatural or mystical as a causal principle. The supernatural does not seem to require that action occur by contact through identifiable causal chains. Supernatural forces from afar or from nowhere were thought to affect one's health and destiny. For instance, in lower Bengal, an area in which cholera was endemic, inhabitants once thought the disease to be controlled by a goddess whom they worshipped to ward off epidemics.[7]

The modern view of action by contact reveals chains of spatial connectivity which can be used to extend our knowledge of other processes. Understanding how cholera is transmitted, for instance, can help us identify links in other processes in other geographic scales. Mapping the dates and locations of cholera epidemics enables us to trace the movements of people and to construct transportation and migration networks among places.

Three great epidemics of cholera occurred in the United States in 1832, 1849 and 1866. By mapping the dates of recorded outbreaks during these years we can retrace the spread of the disease, and by comparing differences in geographic patterns of the epidemic over the three time periods we can infer what were the dominant communication networks within the urban system of the country for those years, and how such a network changed. During the earlier years when the transportation system in the United States was in a rudimentary form the disease spread primarily through spatial proximity. That is, the places that received the disease at a particular time were the ones physically close to the places already containing the disease. In later years, however, the disease was carried along by more developed and integrated links in the urban network so that it spread quickly to other larger cities far away even before it spread to smaller towns and hamlets nearby.[8]

Hence, changes in transportation or paths by which causal agents are transmitted effectively alter the significance of distance and location. When paths follow straight lines and the process is non-hierarchical, cities and towns physically near each other may be causally near. But when the routes are circuitous, and large but distant cities are intensively connected to one another, as Chicago and New York are, the smaller towns physically near such cities, in effect, may be far away or isolated. The significance of physical location and distance depends then on existing channels of communication, and these in turn depend on the present distribution and patterns of things in space.

So familiar are we with action by contact and the interconnection between pattern and process that these principles affect every realm of social action. Proof or innocence in a crime, for example, rests on whether such links in space and time can be established between the accused and the victim. Not knowing these links is tantamount to a mystery or an unsolved case. If, after the end of a motion picture film, the lights in the cinema are turned on to reveal a man who is slumped in his seat, dead from strangulation, and if the theatre receptionist remembered the man walking under his own power into the theatre, we can suppose that the murderer must have been in the theatre at the time the picture was shown. Hence, an airtight alibi for someone would be to have witnesses who could swear the person was away from the theatre at the time. This is the simple but direct application of the principle – action by contact.

If we do not accept the principle, we enter into realms of thought which can only be described as mythical – magical. In science fiction we deliberately enter such a realm by often suspending our scientific conceptions of the role of space in causality. For instance, the murderer of the man in the theatre may never have been in the theatre, and yet committed the murder through some form of thought control over the mind of an unwitting accomplice or through a form of sympathetic magic, such as taking an image of the man in the form of a doll and strangling it.

Although such an explanation would be laughed out of a court of law, it is important to recognise how so much of what we think is normal or strange depends on how the role of space is interpreted. When a different form of thought than the modern Western view prevails, such as the supernatural or magical – mythical, the 'normal' and 'abnormal' are to a great extent altered because of the differences in the role of space in causality. Few of our current legal relationships and social deductions would be self-evident or make sense if we were to adopt a magical view. And yet, such views once had enormous impact, not only in the conduct of social relations, but on the use of the land and the evironment.

For most of the actions and activities studied by the social sciences we have yet to discover as clearly defined a causal agent as the *vibrio cholerae*. This means that we do not possess a well-developed science of human behaviour. But we do not expect things simply to happen, or to appear and disappear, or to be in a place without some reason. We ask where will something be, how and why things occur where they do, and why they occur here, rather than there. We expect the spatial properties of things to affect behaviour and we formulate answers for these questions which are as close as possible to scientific ones and which conform to the principles of causality. Yet, we frequently have only vague ideas about which factors influence behaviour. The degree to which we are unclear about the causes is the degree to which the significant spatial relations are unknown, or known incompletely.

In such cases the application of the scientific perspective may lead us to an image or preconception of the effects of space which comes from a literal interpretation of action by contact, one which is based on the assumption that interaction or influence between two objects is greater if the objects are closer in physical space. Whether such a preconception is useful or misleading, that is, whether or not the interpretation of action by contact is too literal, depends on whether in each particular case it is borne out by the evidence. But proof in social science is difficult to obtain, and without it, one of the dangers is that this image may be taken as the actual situation.

For example, we expect that the attitudes and values of children are influenced by those of their family and friends. Yet we cannot specify what kinds of family contacts would influence children and hence the important locational arrangements of family members are equally difficult to specify (that is, we do not know if being in sufficient contact with parents means being in the same place). Despite our ignorance of such matters there is a conviction that being close to parents means being under their influence. Such a conviction stems in part from a perhaps too literal interpretation of action by contact as physical proximity.

A too literal interpretation of the significance of spatial relations often occurs in social science and in the practical realm. We assume that certain places, shapes and distances are important. However, even if they are not, or even if the significance is unsubstantiated, our beliefs come to invest such places and configurations with import. For instance, we believe there is a right and wrong side of the tracks; we make political choices based on the region from which a candidate comes; national political tickets are often geographically balanced; we explain a person's behaviour by saying he is a New Englander, or she is from Southern California; we suppose that a change of scene will do one good; and we believe in the domino theory (that is, that communism will most likely spread to places near rather than to places far). While the effect of actually being in such places or having events in such configurations has not been clearly established, they nevertheless are used to explain and predict behaviour.

Even though it is often misapplied, action by contact is the basis of the conception of space in social sciences and in our practical view. Action by contact is a very general principle and does not by itself disclose the particular chains and configurations through which interactions occur. When we search for these and attempt to generalise about them – when we consider how in general particular kinds of spatial configurations affect human behaviour, how in general places develop, how in general physical separation in space affects interactions – we are no longer looking at things through both the practical realm and the social scientific but rather we are examining the problem primarily from the latter. Moreover, in their zeal to find such generalisations, social scientists often overlook the role which action by contact must play. We turn now to some of the problems that the social sciences, and especially geography, have encountered in the explorations of particular paths and networks of interaction in their attempts to find the specific relationships between space and substance. We intend by this discussion to increase our awareness of the problems of interpreting action by contact in the human realm and thus of the problems of conceptually linking space and substance.

Significance of space

The (logically) first conceptual link between things or substances and space occurs through locating the former in space. This provides a means for describing and individuating facts. The second conceptual link occurs when we establish the significance of the spatial properties of things in scientific laws and theories.

Both uses of space are of interest to geography. Besides its well-known task of keeping track of the location of things, geography analyses the manifold physical spatial relations among activities on the surface of the earth in order to attain a scientific understanding of the significance of space for human behaviour.

Such a quest for significance makes us re-examine the traditional saying that 'things occur in space' by asking *where, how* and *why* they do occur where they do. Such questions take the original fusion of space and things and separate them into two categories. On the spatial side there are concepts such as location, distance and shapes, while on the thing (or substance) side there are the innumerable categories reflected in our terms for 'things'. Emphasis on the spatial side has stimulated spatial questions and the spatial perspective. Emphasis on the other, as we shall see, has led to the systematic social sciences and on occasion to the underemphasis on spatial relations. A clear understanding of the conceptions of space in social science is required to evaluate the contribution of the two approaches, and to determine how far and in what way they can be separated and still be discussed in a scientifically meaningful way, and how they can be recombined to provide an integrated view of human behaviour.

Geography seeks answers about the effects of properties of space on behaviour through the methods of science. Science in general involves separating what are thought to be significant factors and determining their possible causal interconnections. Ideally, such separation would occur in a controlled laboratory environment. But for much of the study of human behaviour this is impossible. In such cases the separation of things is primarily conceptual. We may ask people to isolate and evaluate the elements in an environment that they think are important or that they believe contribute to their behaviour, or we may look for recurring situations and explore similarities and differences among the elements and their interrelationships.

The same applies to understanding the significance of spatial relations in behaviour. Since things are in space and have spatial properties, we cannot actually separate things from space. All we can do is alter their spatial arrangements, find similar things in different spatial configurations, or have people imagine such alterations in order to determine the effects of space on behaviour. We may ask people if and why they would prefer *x* over here rather than there, or we may investigate such recurring relationships as transportation networks, population density and land-use patterns to seek explanations about why things occur where they do, or what effect the distances of the networks have on land use. We can, of course, increase the number of variables examined until we have models of entire landscapes.

In practice the social sciences' approach has encountered such enormously complex empirical problems that much of the effort of the disciplines has been devoted to the development of elaborate statistical methods for avoiding them and for determining causal relations in non-experimental studies.[9] The application of the scientific method to the quest for the significance of spatial relations has led to particular needs in statistical methods and research procedures.[10] But also it has led to the need for a clearer understanding of the social science conception of space, one which would help address such problems as the manner and extent to which we can expect to conceptually isolate space and its properties and determine their effects on behaviour in a scientifically significant way. This is a difficult problem, for space is manifested through things. Without things or substances, space has no embodiment and, from the conflation of action by contact and the conservation of energy, no effect.

To complicate matters, people's behaviour is affected by the way they think things are, as well as by the way things actually are. These thoughts may differ greatly from reality and involve both different descriptions and evaluations of space. Whether the domino theory works or whether particular alliances of encirclement contain the enemy, whether a place actually possesses certain characteristics, or whether two things are really close, people may believe those things to be so, and behave accordingly. If the beliefs do not conform to the facts, then, presumably, action based on such beliefs may not succeed. But it is likely that the actions will be undertaken according to what things seem to be like.

Determining how people believe things to be is as much a factual matter as determining what things are like. Nonetheless, it is a difficult task, and part of our beliefs may be misconceptions of a social science. We have already seen that a too literal interpretation of the scientific meaning of space may alter our beliefs about space and place and affect our behaviour, thus creating new attitudes for a social science to explore. There are other impediments in the way of a social science analysis of space and some of these can be related to the complexities resulting from the application of science.

The wide acceptance of the scientific approach and its offspring, technology, has created its own cleavages between space and thing, and has made their recombination difficult. At the present, many decisions are so complex and rely so heavily on technology and social institutions that we can rarely describe the chains of influences in space or time which our decisions involve. If we need to talk with someone, all we may have to do is pick up a telephone and dial a number. We need little information beyond knowing the number, where the telephone is, and how to get to it. We do not need to know where the other conversant is, or through which lines the message has been routed, or where the energy for transmitting the message comes from. Nor do we need to know anything about the telephone company. All of these matters are taken care of for us. As a result, our awareness of the processes and the spatial manifestations which our actions involve becomes fragmentary and constricted.

Through the process of making distant places accessible, we lose sight of the spatial consequences of our actions. Much of social science research is directed at reconstructing these chains of influence so that we will know what parts of the society are interrelated. This loss of sight of the spatial consequences of actions does not apply to every action or to everyone equally. In person-to-person contact we can create and control our channels of communication and can see what effects we have on others, but we cannot when we use the telephone. Telephone communication has behind it employees who create and service such channels so that others will never have to worry about them.

The prevalence of technology and the division of labour, which have so complicated our activities and fragmented our responsibilities, have led us to think of decisions and actions in terms of their degree of connection with space. An action is thought to be less spatial or even non-spatial, if we do not know of its spatial consequences or manifestations at all or in a geographic scale in which we are interested. Firing a bullet at a target may well be conceived of as a spatial act, talking on the phone and whistling in the wind as less spatial, and having an idea as simply ethereal.

Prying things from space, and not entirely recombining them, is reflected in, and further rigidified by, our political and social institutions. Hierarchical political territories like cities, counties, states, nations, and so on, fragment and obscure the details of the links between behaviour and space. Geographic manifestations of behaviour may be clear at only one of several levels or scales. From the point of view of the other levels, the behaviour appears less spatial or non-spatial. This is seen, for instance, in national legislation. Some legislation and decision-making may have a predetermined impact on only the national level, others on the state levels or the cities, and so on, and some may not be specifiable at all. But regardless at what level or for what location the impacts may be specified, there always remain more levels and locations to be worked out.

As though to make up for this fragmentation, there are agencies of government which emphasise the 'regional' or 'local' impact of actions, or those parts of the causal chain which touch only their particular areas. These include state and local governments, regional planning commissions, school boards and zoning commissions. The concerns of these units are often thought to be pre-eminently spatial, while in fact they are no more or less spatial than those of other jurisdictions. They are simply local. Moreover, their interests may be too circumscribed to evaluate and control properly processes that affect their areas or to contain spatially the consequences of their own initiations. We can zone an area for single-family dwelling units with the unplanned consequence of forcing apartments and businesses to locate elsewhere, or we can dispose of waste in an administrative unit to the satisfaction of its inhabitants with the unplanned result that it interferes with broader natural and social systems elsewhere. In short, the boundaries of such areal units may not coincide with the boundaries of physical and social systems. Yet there are innumerable units like these at all scales. They are in part created to regain control and knowledge of the geographic impact of actions at different scales. But they also further segment and complicate causal chains and make space and human action more difficult to connect.

A division of space from things occurs in yet another way through the effects of the academic disciplines. Geography, regional planning, landscape architecture, regional science, land economics and rural sociology are among the disciplines that are concerned to a great extent with spatial relations, and there are the rest of the social sciences which profess little or no concern with the spatial manifestations of actions. Some studies in the latter even obscure the spatial properties of data by aggregation, by pretending that they occurred nowhere, or by disguising the location.[11]

A distillation of these several cleavages between space and things can be sensed in qualifications which often attend the statements that things occur in space. While things are admitted to occur in space, perhaps not all things do. Some argue that mental states such as attitudes, values and ideas do not exist in space and do not have spatial properties. Such facts, it is claimed, are different from material objects and things. They may not be things in the normal sense of the word. To this it could be countered that for us to know about such states in others, we must observe their physical manifestations and hence their corresponding spatial properties. Certainly the definitions of such mental states have been vague and hence their lack of clear spatial manifestations may be due to our inability to recognise an instance of them.[12]

Some may argue that there seems to be a greater spatial connotation to the term object than to the term event and that the latter seems more temporal than the former. Again, a possible rejoinder is that the difference may be simply one of emphasising different characteristics which both possess. Perhaps events are fast-moving objects, while objects are slow-moving events. A continent, for instance, may seem to be an enormous physical object. But in the perspective of geological time continents would appear as a sequence of events, a series of objects connected over time.[13]

Disciplinary divisions, political units and technology are examples of the many areas in which modern society strains the connection between space and thing and presents problems which a social science analysis of the effects of space on behaviour must address. Extending our awareness to other modes of thought adds dramatically to the differences in the ways space and things are linked and space conceived. These other conceptions of space are not only different in the abstract, they have also affected the way in which man uses the land. Some of the differences can be suggested by the way in which a single shape, such as the circle, is evaluated in different modes.

Space in non-scientific modes

From a scientific perspective the circle is the shape which encloses the greatest area with the smallest perimeter. Hence, it is the most compact. It is this property of the shape which makes it an efficient settlement form. A circular settlement allows both accessibility to the centre and efficient defensive perimeters. But this shape may be valued for reasons other than compactness. We have already seen that, in a mythical–magical view, the use of space is very different from its use in most realms of science. Things from afar can affect one another without intervening substances. That is, action can occur at-a-distance.[14] Another characteristic is that a spatial form or shape, as a symbol of something, may be thought to possess the powers of the thing it represents. If the cosmos is thought to be round, creating a circle will not only symbolise the cosmos but in effect reproduce it in microcosm. Therefore, settlements having circular shapes may, according to this view, be the best design because they would tap the cosmic forces and be in sympathy with the heavens. The Hottentots designed their huts, their kraals and their villages in circles. The chief's hut is always located on the part of the circle in line with the rising sun. So closely was this followed that from its location one can tell at what season the camp was settled.[15]

Space and its properties are conceived differently in the realms of art. Spatial configurations and shapes are very powerful images in poetry. The circle especially was important in Elizabethan poetry when it was thought to exemplify divine order, but it finds a place in

the imagery of all ages.[16] And each time it may possess different
poetic meanings. For instance, it has been an image of chaos:

> The world's a seeming paradise, but her own
> And man's tormentor;
> Appearing fix'd, yet but a rolling stone
> Without a tenter;
> It is a vast circumference, where none
> Can find a centre.[17]

And it has been an image of order:

> He took the gold'n compasses, prepar'd
> In God's Eternal store, to circumscribe
> This Universe, and all created things.
> One foot he *centerd*, and the other turn'd
> Round through the vast profunditie obscure,
> And said, 'thus farr extend, thus farr thy bounds,
> This be thy just Circumference, O World.'[18]

In the visual arts (painting, sculpture and architecture), space and its
properties are described in very different terms than in science. The
juxtaposition of shapes and forms are said to possess tensions and
resolutions, harmonies and discordances, balances and disequilib-
riums. Space can be bounded and unbounded, static and dynamic. In
art based on circular motifs, the interpretation of the concentric
spatial forms involves such antitheses as core – periphery, inward –
outward, bounded – unbounded, infinite – finite, focused – unfo-
cused. According to artists José and Mariam Argüelles, 'the purest,
simplest, yet most encompassing form is the circle . . . .'[19]

Spatial imagery is powerful and versatile in other realms of
thought. For example, the circle has been an important image in
metaphysics. Contemplating mandalas has long been held by Eastern
religions to lead to spiritual enlightenment. Figure 1.1 is an example
of a mandala found in Tibetan Buddhism.

The dynamic and multiple relationships between the centre of a
circle or sphere, its circumference, and the area within, have been
used as images of the Christian trinity as well. For example, Nicolas
Cusanus, a fifteenth-century theologian, 'has the Father, as the
generative principle, hold the center, from which the Son issues as a

FIGURE 1.1   *Yantra form of mandala*

power equal in kind to that of God. The Holy Ghost unifies the two and closes the whole by the circumference.' According to seventeenth-century scientist Johann Kepler, though, 'the image of the triune God is in the spherical surface, that is to say, the Father is in the center, the Son is in the outer surface, and the Holy Ghost is in the equality of relation between point and circumference.'[20]

Whereas the same geometric form may be construed differently, the same or similar patterns of relationships can be represented by different geometric forms. For instance, the triune can also be symbolised by the equilateral triangle; each vertex is equally prominent and situated with respect to the others. Circumscribing the triangle ⊕ directs our thoughts along the circumference from vertex to vertex.

If we turn from example of shapes and forms to attitudes about places and territories we find differences in the conceptions of space determined by different social orders. The imprimatur of twentieth-century power and authority on the landscape is the territorial state and its innumerable areal subdivisions. The intangibles of social order and political authority in a complex state are reified, on the ground, by political boundaries, and these completely partition the inhabitable land surface of the earth. In capitalist systems, segmentation of terrestrial space occurs through the private ownership of land as well. This institution allows land speculation in which a parcel of land may not be valued in terms of its present content but rather in terms of future possibilities. Until the price is right, it allows land to be held as though it were empty, and prevents activities and things from being located there.

Both private ownership of land and the territorial state contrast with societal views of space in primitive societies. In the primitive view, land is not a thing that can be cut into pieces and sold as parcels. Land is not a piece of space within a larger spatial system. On the contrary, it is seen in terms of social relations.[21] The people, as part of nature, are intimately linked to the land. To belong to a territory or place is a social concept which requires first and foremost belonging to a societal unit. The land itself is in the possession of the group as a whole. It is not privately partitioned and owned. Moreover, it is alive with the spirits and history of the people, and places on it are sacred.

Such attitudes to place are markedly different from those found in the industrialised West and have led to misunderstandings between the two societies, as in the 'clouded titles' by which the white man has claimed the land of the 'savages'.[22]

These examples give us an idea of how places, territories, shapes and other spatial relations are evaluated differently through different perspectives and modes of thought. All individuals and groups use more than one mode of thought to evaluate space and its properties, and even to evaluate a single form, shape or place. Although some modes are more distinct and developed in one society than in another, all normal human beings have the potential to express themselves in each of the modes. Intrinsic within human thought is the potential for each mode.

Until recently most of the geographic inquiry about space has concerned the meanings of space in the scientific mode of thought and their applicability to human behaviour.[23] Among the results of this research have been concepts of space and place such as those found in location theory, chorology, and cognitive and mental maps. These, as we shall see, form the core of a social science conception of space. But their differences are subtle when compared with the ranges of meaning that arise from non-scientific conceptions. As it becomes more evident that other conceptions of space affect behaviour and attitudes towards the earth, and that these are not (yet) embraced by the social sciences, it becomes essential that we attempt to expand our analysis and define, as clearly as possible, the range of conceptions of space and their interconnections which have directed man's behaviour towards the landscape. This requires a general framework to help us explore the interrelationship of conceptions of space in modes of thought.

Framework

We have already suggested that addressing conceptions of space leads to a division between space and things or substances. How the two are conceptually separated and recombined varies from mode to mode and is a sensitive indicator of the differences in conceptions of space. If we consider the range of modes of thought that should be addressed by such a framework (that is, the scientific, the social-scientific, the aesthetic, the child's view, the practical, the mythical–magical and the societal) it becomes clear that the relationship between space and substance that characterises each of them depends on the degree of what we shall here roughly refer to as the subjectivity or objectivity of the modes and of the facts they explore. For instance, space in art is seen differently than it is in science primarily because of the more subjective nature of art. Art's primary focus is in subjectively symbolising subjective facts such as feelings and emotions, while science attempts to objectively analyse objective facts. Different mixtures of subjectivity and objectivity characterise the other modes of thought as well. Therefore, subjectivity and objectivity (in terms of both subject matter and method of analysis), in addition to space and substance, will be used as the basis of our framework. Employing the subjective–objective distinction is supported further by the fact that we are analysing conceptions of space from a geographic perspective. This means that the scientific meanings will be the benchmarks of our analysis, and science is assumed to be the most objective view while non-scientific views are held by science to be, to varying degrees, subjective. It should also be remembered that the framework considers only those aspects of the modes of thought which are significant for understanding conceptions of space of interest to geography. It is not intended as an analysis of all possible modes of thought in all of their richness and interrelationships.

We will use then the distinctions between subjective–objective, space–substance as the basis for the framework. The framework has two intersecting axes or dimensions forming a conceptual surface on which the meanings of space can be placed; an intellectual area, if you will, for conceptions of space. A schematic of the framework is offered in Figure 1.2. The end points of the vertical axis are space and substance, and the end points of the horizontal axis are subjectivity and objectivity, referring to both the subject matter and to the method of analysis. In this regard there are really two parallel horizontal axes collapsed into one. The drawing should not, however, be taken too literally. First, it is difficult if not impossible to isolate pure objectivity and subjectivity so that the objective end of the axis should be thought of as containing elements of the subjective and *vice versa*. Second, contrary to the rigidity of the diagram, the conceptual surface is dynamic and fluid. These axes are created by the conceptual separation of properties of reality which, when seen naively, are connected or fused. The more abstract the thoughts, the greater the separation, the longer the axes, and the greater the extent of the surface.

Because of the limited knowledge of the social sciences, our

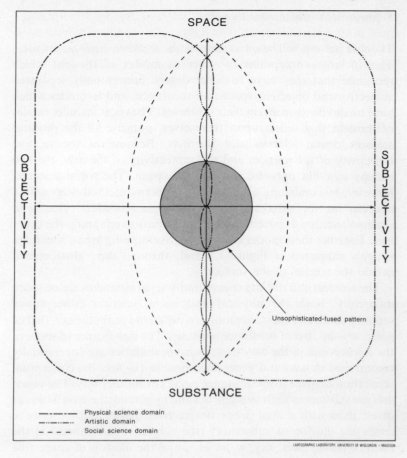

SPACE

OBJECTIVITY

SUBJECTIVITY

Unsophisticated-fused pattern

SUBSTANCE

—·—·— Physical science domain

······⋯···· Artistic domain

— — — — Social science domain

CARTOGRAPHIC LABORATORY, UNIVERSITY OF WISCONSIN – MADISON

FIGURE 1.2 *Conceptual surface: sophisticated–fragmented pattern*

understanding of these extensions, their degree of separation, and the links between them is limited, fragmented, and so too is our knowledge of the surface. We know little about its contours, about whether it is flat, curved or convoluted, where precisely on it the other conceptions are projected, and how the surface itself may appear altered as we submerge ourselves in one or another mode. Despite these limitations, it is still possible to discern the outlines of two very different patterns of thought which can be projected on to this surface, and within which the modes and their conceptions of space can be located and categorised.

Sophisticated–fragmented thought

The first pattern will be referred to as the *sophisticated–fragmented*. This includes conceptions of space in modes of thought which recognise that they have to some degree conceptually separated subjective and objective, space and substance, and to modes which have made efforts to attain their syntheses. That is, it includes modes of thought that reflect upon themselves, namely, all the physical sciences, social sciences and the arts. Because of the relative objectivity of the sciences and the subjectivity of the arts, the two occupy vast but opposite sides of the space. The social sciences, however, by combining a supposedly objective methodology with a concern for the subjective, as in feelings, attitudes, values and illusions, occupy a space between and in part overlapping the other two. Together these modes cover large, overlapping areas, which are roughly suggested in Figure 1.1 and, through their abstractions, extend the reaches of the surface.

The sciences and the arts conceptually separate and evaluate space differently. Both the physical and social sciences conceptually separate space and substance into two different vocabularies – that of geometry and that of substance or things. The significance of space in the sciences lies in the way these two vocabularies are conceptually recombined in laws and generalisations. In the arts the conceptual separation of space from substance occurs in another way. The visual arts use substances such as pigments and by placing them on a canvas invest them with spatial properties such as shapes and patterns to create the illusion of substances (the subjective substances) in the form of rocks, trees, sky, water etc., and the illusion of space (the subjective space) in the form of a vista, or place or landscape. From the artistic fusion of space and thing comes the illusion of their separation.

In the non-visual arts such as poetry and prose, words are used to create the illusion of substance and place or space. In both the visual and non-visual arts this illusion of space is its conceptual separation. The meaning or significance of this illusion depends on the feelings it symbolises. The illusion of space may symbolise feelings about space itself, as when the landscape may symbolise vastness or confinement, or the illusion of space may symbolise feelings which need not be associated with space itself, as when either the landscape or a portrait (which also creates an illusion of place or scene) symbolises happiness or sadness. Whereas the significance of space in science is determined by the conceptual recombination of space with substance in laws, the significance of space in art lies in its connection to feeling, that is, in the import of the illusion.

While the characteristics of the surface and the two axes of the framework may be conceived differently within each perspective, they all have in common the conceptual 'capacity' for such a surface. They attempt to separate conceptually the subjective from the objective, the space from the substance, in order to represent them and their interconnections in terms of their own logic and symbolic forms. They are sophisticated because in so doing, they handle their concepts maturely by not confusing the symbols with what they represent. The scientist realises that the manipulations of his equations are not going to affect the things they represent, and the artist realises that painting a landscape symbolic of feelings such as tranquillity or turbulence will in no way alter the nature of the landscape.

But the modes are fragmented because they do not encompass the entire surface and do not completely recombine the elements of analysis they have conceptually separated. Moreover, the concepts of space become specialised in each mode: physical science conceiving of it in terms of a geometric system; social science in part sharing this conception but altering it through the introduction of feelings and perceptions in the forms of specific places and cognitive and perceptual spaces; and art creating semblances of space and shapes in virtual space.[24]

Unsophisticated–fused thought

The second pattern of thought will be called the *unsophisticated–fused*. It embraces those conceptions of space in modes of thought in which the conceptual separations of space, substance, objective and subjective do not occur at a high level of abstraction, if at all. In fact, as a group these modes are unaware that there could be such a conceptual surface in which to examine symbolic categories. This group includes the child's view, the practical view and the mythical–magical view. Figure 1.3 is a diagram of the conceptual surface from the viewpoint of these modes. The smaller extent of the surface and the dotted lines indicate that these modes are unaware of what little conceptual separation they employ. The modes are located around

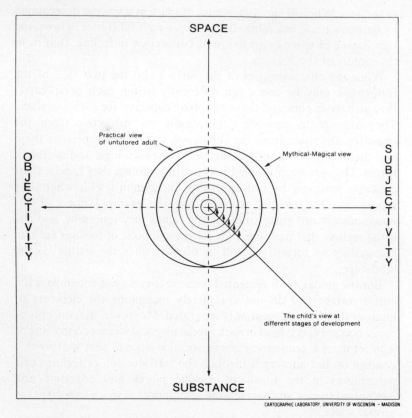

FIGURE 1.3   *Conceptual surface: unsophisticated–fused pattern*

the origin of the surface and occupy smaller areas than do the sophisticated–fragmented. The solid parts of the axes extending beyond the circles refer to the normal extent of the surfaces as viewed from the sophisticated–fragmented patterns.

These modes are unsophisticated because of their low levels of abstraction and their frequent confusion or conflation of symbols with what they represent, as when the child or a believer in magic destroys a symbol of a thing, and expects the thing itself to disappear, or creates a symbol and expects to create its referent, as in a rain dance to bring on the rain.

These modes are fused because their symbols are neither highly abstracted nor primarily directed along one of the axes as in the sophisticated–fragmented. Space is not conceptually far removed from the substance or things which take form. Their symbols or concepts often encompass both facts and feelings. These modes do not include a distinct geometry, or science or art. Although actions and activities within these modes may be artistic, scientific and practical, they are rarely only one of these and are not compartmentalised into these categories.

The societal conception incorporates aspects of all of the modes in both patterns. Hence it is omitted from the diagrams of the conceptual surface.

The major differences between the two patterns and among the modes of thought is in their use of symbols. The most fundamental reason for the fragmentation of the surface among the sophisticated modes and also for the peculiar, and so far, unsuccessful attempts of the social sciences to analyse feelings and emotions, is the fundamentally different way in which facts and feelings seem to be symbolised, especially in the modes of thought which have specialised in such symbolism–namely, science and art. Generally, the sciences can use symbols to represent facts, and the symbols are conventionally defined. The symbolic vehicles themselves have no meaning (within science) apart from the conventional definition assigned to them. Hence, an x in science will mean what we define it to be, and there is nothing intrinsically similar between the physical shape and appearance of the x and what we use it to represent. We could just as easily use any other letter in the alphabet or any other sign. But once we have agreed upon the x, its meaning is relatively clear and constant.

On the other hand, feelings are most successfully and naturally expressed by symbols whose meanings seem to be in part affixed to the vehicles themselves. They do not appear to be conventionally or arbitrarily decided. These symbols are said to *present* rather than *represent*.[25] Their meanings, unlike the scientific ones, are not clearly specified and stable, but rather depend far more on the contexts in which they appear. In terms of the objective side, such symbols seem to be over-determined, fluid and context-dependent. Thus ensues the observation that art cannot be reduced to invariant elements with constant meanings, but rather has to be interpreted each time anew in terms of the whole. Accordingly, there is no language of art, no vocabulary which can be combined by syntactical rules to symbolise or express specific meanings. Rather, the elements of an artistic

creation depend on the whole for their identification and meaning.
The basis for the unsophisticated–fused pattern is the way in which
symbols are seen and used. In these modes, the world is seen so
intensely and closely that feelings are often projected on to it,
animating the world. The symbols only partially and tentatively
separate the subjective from the objective, because they themselves
are seen as sharing properties of the things they represent. Hence the
meanings of these symbols do not seem to be arbitrarily defined.
Unlike the symbols in art which do seem to share properties of their
referents, the symbols in the unsophisticated–fused view are easily
confused with what they represent. It remains to be seen whether
these differences in symbolisation are real in the sense that they are
due to fundamentally different logical relationships between symbol
and referents, or are apparent in that they seem to be different from a
scientific perspective because of our scientifically imperfect understanding of feelings and their symbols.[26]

Order of topics

The sophisticated–fragmented and the unsophisticated–fused define
two general patterns in the conceptions of space. These patterns will
form the respective headings of Parts II and III of this book. The two
patterns and the two parts, moreover, are in dependent opposition to
one another. The unsophisticated–fused view may be thought of as
the simpler, more basic and more primordial view of space, but it is
unaware of its own characteristics and of its relationship to other
views. Such an awareness comes only from the sophisticated–
fragmented view. Although providing a perspective, the sophisticated–fragmented view has thus far been unable to capture to any
great extent and without distortion the unsophisticated–fused
conceptions.

Often it is well to begin with the primordial. We will start with the
sophisticated–fragmented, however, because this work is not intended as a history of the development of the conceptions of space.
Rather it is an exploration of the issues about space in social thought
from the perspectives of a science and a social science, namely,
geography. Our inquiry and problems, therefore, are posed by this
point of view and by its conceptions of space and we will examine the
simpler view in light of it.

After a more detailed view in Chapter 2 of the scientific perspective and the relationship between the objective and the subjective, we will begin our journey on the conceptual surface in Chapter 3 from the objective side, exploring traditional social science views of space, and enter more deeply into the realm of the subjective in Chapter 4 by considering the behavioural approach and conceptions of space in art. Chapters 3 and 4 form the basis of the sophisticated–fragmented pattern. In Part III the surface will be contracted and axes blurred as we turn to the meanings of space in the unsophisticated–fused modes. We begin in Chapter 5 with the most basic, the child's view of space, which in maturity, forms the standard practical conception of space. In Chapter 6 we increase the degree of abstraction, by considering the conception of space in myth and magic. This view is based in part on the child's conception and on the practical view of space. In part IV, Chapter 7, we will explore the reformulations and syntheses of the two major patterns at the societal level, first in primitive societies and then in civilised ones. In addition to synthesising the other views of space, the societal view has had a great influence on the development of science and social science and has, perhaps more than any other, left a visible and permanent impact on the earth's surface. The implications of our analysis for the social science disciplines will be discussed in Chapter 8.

Before we begin we should recall that geography is not only about the nature of space and spatial relations but also about the nature of man and his environment. The reader may wish to bear in mind that with some shifts in emphasis these axes can be used as a framework for this perspective as well. For example, the space – substance axis could be collapsed and the subjective–objective axis unpacked into two, one representing the distinction between subject matter (with the subjective end of the axis labelled self or man and the objective end labelled world or nature) and the other representing the distinction in types of analyses. The conceptual surface formed by these two axes could then be used to locate modes of thought and their conceptions of man-nature.

Part II
Sophisticated–fragmented Patterns

2
Science and Subjectivity

Scientific knowledge has come to epitomise what is meant by studying things objectively. The success of its predictions and the scope of its explanations have given us confidence that the symbolic form of science–its concepts, laws and theories–accurately reflects the facts and processes in the physical world. Philosophers of science under the influence of logical positivism attempted to make the objective elements of science explicit by looking for general logical principles for the selection and verification of scientific statements. The search produced the complex thesis that there is a neutral, objective, value-free language in which the facts of the world can be described and with which the statements of science can be evaluated. The terms of science are to refer ultimately to the facts we sense and experience and which are describable in this value-free language. Apart from logical terms, sciences are to include only such empirical terms. Moreover, there are logical criteria that can be used to decide what laws or theories best explain a set of facts and relationships.

Confidence in this thesis has been shaken by recent works in the history and philosophy of science.[1] A realist view has shown that just as our perceptions are influenced by our conceptions, so too are scientific descriptions and explanations of the facts of the world phrased in a language which, while appearing to be objective, is nevertheless influenced by the scientific theories purporting to explain the facts. Also the selection of one theory over another historically has not been made on the basis of a clear set of logical principles that would apply each time there was a choice among competing theories. Moreover, the positivist conception of, and emphasis on, empirical terms or concepts does not reflect the way scientists actually conceive of and use laws and theories. It especially does not take account of the attention that scientists pay to the hidden and 'non-observable' forces and structures which are thought to direct and control the facts we observe and which are included in the sciences as general assumptions and as referents of theoretical terms. Hence the positivist characterisation of laws and theories has tended to make science appear to be concerned primarily with the apparent or even epiphenomenal elements of the world.

To hold, as do realists, that there may not be a logical procedure to ensure objectivity and that elements of judgment and context enter into scientific description and evaluation does not mean that science is subjective. To so argue would be to throw out the baby with the bath water – to despair of characterising the actual differences between science and other modes such as art or myth and magic. These criticisms of positivism do mean that actual science is not entirely objective and that positivism offers an important but incomplete characterisation of the way science works. Despite the validity of the realist criticisms there is as yet neither a suitable amendment to positivism which would embrace these criticisms nor a comprehensive realist formulation which itself is not (equally?) problematical. Given this uncertain state of the characterisation of the scientific enterprise, we shall look for common ground among the two views which will allow us to see how science can attain a relative objectivity and use this as the basis for a minimum framework on which to build a more comprehensive realist view. Our analysis will be presented as realist amendments to positivistic characterisations of concepts, laws and theories.

Scientific objectivity, concepts and symbols

Considering then the relativity of the terms objective and subjective, we see that each end of our objective–subjective axis contains elements of the other. For the sake of brevity we shall not always use 'relative' when talking about objectivity and subjectivity although we always mean it to accompany the terms. In order to be more specific about the properties of this axis we must distinguish between the two meanings of objective and subjective which are incorporated in the axis and which are often confused.[2] The first meaning refers to the subject-matter studied and the second refers to how the subject is studied. With regard to the subject-matter, we can study objective facts such as rocks and streams, or we can study subjective facts such as feelings and emotions. The physical sciences in general have studied objective facts, the social sciences have studied both. In the method of study, objectivity is associated primarily with science and perhaps social science, and subjectivity is associated primarily with the arts. Since our framework emphasises the scientific perspective, we will consider first the problem of objectively studying objective facts.

Accurate, orderly and objective description of things is part of science. Pure description may seem to require no other talents than patience and a lack of imagination, but even the simplest description involves a complex association of the general with the specific, that is, of recognising class and instances. To describe a glass as a transparent cylinder is to use at least two empirical concepts or terms. The description presupposes a knowledge of the meanings of 'transparent' and 'cylinder' and a recognition that the object being described is in fact an instance of them. These presuppositions are involved in both scientific and non-scientific description. A difference between science and natural or ordinary language is that scientific descriptions are more exact because the terms or concepts are more precisely and explicitly defined. The relatively clear and precise definition of the terms, or concepts of science gives them a meaning which we will refer to as a *meaning*$_1$.[3]

Definitions in science, as in natural language, are usually in terms of other terms which have a meaning$_1$. The physical science definition of force is the acceleration a body with a given weight experiences along the horizontal without effects of friction. The definition of acceleration is the rate of change of its velocity with time. In order to avoid complete circularity, some terms of science are undefined within the vocabulary of science. These are called *primitive* terms and their meanings are derived through such procedures as *ostensive* definitions (that is, pointing), as when we point to a particular platinum cylinder which is preserved in the International Bureau of Weights and Measures near Paris, and say, 'This is a kilogram.' The non-ostensive scientific definitions range in complexity. Some require not only terms to define them, but operations that must be performed in order for the observer to see what is being defined. These are called *operational definitions*. Electrical current along a wire is defined operationally as the amount a needle on an *ammeter* registers when the electrodes of the meter are placed along the wire.

Scientific meanings$_1$ are matters of convention, made explicit by the scientific investigators and agreed upon by the scientific community. A term can mean anything one chooses. We could have decided upon another object as the definition of a kilogram. But the condition of there being instances of the term or concepts, that there actually be something to which the concept refers, is a matter of fact and is not guaranteed by the clarity of the concept. There may be clearly defined concepts which have no instances, that is, which do not refer to anything that actually exists. Although we know precisely what we $mean_1$ by a fifty-foot human being, there are no such creatures. Concepts which refer to things – which have instances – are called *empirical terms*. These are distinguished from another and much smaller class of terms which refer to logical relationships. These are called *logical terms* and refer to such relationships as 'and', 'or', 'all', 'either'. There are in addition the extremely important *theoretical terms* which do not have instances in the sense that empirical terms have. These occur in the more advanced stages of science and provide a context for the theory while these terms are not about the directly 'observable' world, they are none the less about reality.

When we say that there is an *instance* of a concept we are both describing and classifying something. The sentence, 'Irving is intelligent', states that Irving is an instance of the concept intelligence. Certainly any thing may be an instance of many concepts. A glass may be a container, a piece of silicon, a work of art. To say what it is scientifically, is to subsume it under one or more scientific concepts. Classification is the often complex procedure by which we subsume facts within concepts and concepts within other concepts. Concepts do not occur in the sense that facts occur, but rather in the sense that the terms and definitions are written down or otherwise evidenced empirically.

The $meaning_1$ of a term is a convention and such conventions are neither right nor wrong. Rather they are significant or not. The significance of a clearly defined concept gives the concept another extremely important meaning which we shall refer to as its $meaning_2$.[4] A concept is *significant* or has a $meaning_2$ when we know that the facts it refers to are empirically associated with other facts, which also are subsumable under relatively well-defined concepts. In other words, significance is knowing that if and when an instance of a concept A occurs, instances of B and C . . . N occur. A, B, C, and so on, are significant concepts. For example, we have a great deal of confidence that if the temperature of water falls below $32°$ Fahrenheit at standard pressure conditions, ice will occur. The concepts water, temperature, pressure and ice are well defined and significant.

We use the term *law* to identify conjunctions of concepts which generalise real relationships among the facts they subsume. In general they include only logical terms like 'and', 'or' and empirical concepts like water and temperature. They do not include proper names, like Richard; specific locations or places, like Madison; or times, like the Renaissance.[5] In other words, they are about the occurrence of things in general and they apply only when the facts they subsume occur.

Knowledge of significant concepts – the establishment of laws – is one of the hallmarks of science for it offers more than accurate orderly description and classification. Theoretically at least, according to the positivist view, it allows for the symmetrical function of explanation and prediction of facts subsumed by the laws. The lawful relationship among water, temperature and pressure can be used to predict the occurrence of ice. Ice will occur when water falls below 32° Fahrenheit, and the law can explain why ice occurred. It occurred because the temperature of the water was below 32° Fahrenheit. In general, if an event *b* is clearly an instance of a concept *B* and we have a law, 'If *A* then *B*', the occurrence of *b* is explained by the law, if it is accompanied or preceded by the occurrences defined in *A*. Similarly, if instances of *A* occur, then *b* will occur. Often we do not appeal directly to the general law but say simply that *a* caused or causes *b*, in which case we assume the generalisation if *A* then *B* is understood.

The examples of generalisations or laws we have given thus far are deterministic, but generalisations or laws in social science are often probabilistic, in which case we should write the connections among the concepts as 'If *A* then probably *B*'. In such laws the explanation and prediction are in terms of the likelihood of the occurrences of the class of events, *B*, although when the probability is close to one we may then discuss the explanation or prediction of an individual event. Furthermore, in probabilistic generalisations the relationship between the generalisation and the event to be explained is inductive while in a non-probabilistic or deterministic generalisation it is deductive. That is, in probabilistic generalisations the facts to be explained are not logically implied by the generalisation and the initial conditions, while in deterministic generalisations they are. For our discussions of spatial relations these distinctions are not important, and to simplify matters, we will use examples of deterministic laws or generalisations with the understanding that what is said or exemplified by them applies also to probabilistic laws or generalisations.

In generalisations or laws, the 'if' clause refers to the antecedent conditions and the 'then' clause to the consequent conditions, but these terms do not always imply that there will be a temporal order to the instances of the concepts, that the a's will precede the b's in time. The instances of the concepts in the laws may occur simultaneously. We will call these *static laws*. In these cases the law can still predict in the sense that by knowing only some of the occurrences we will be able to infer that others which have not yet been observed are also present.[6] Such static laws include the laws of geometry and equilibrium laws. For instance, without measuring the hypotenuse of a right-angle triangle, we can predict its length if we know the lengths of the other two sides; or for a lever we can predict the weight (w_1) of an object on a lever (d_1) distance from the fulcrum if we know the weight (w_2) of the other object and its distance (d_2) from the fulcrum; or we can expect to find certain mental characteristics exhibited by a person who has Downs syndrome.

What is or is not significant (or a law) is extremely important to know and positivists have held that there are two general and related criteria that can be employed to help decide. The first is about the degree to which associations conform to experience, that is, the degree to which they are inductively verified. This criterion places the determination on the shoulders of statistics which is to give an unbiased, objective, measure of the degree of association ranging from a correlation of negative one (perfect negative correlation), to zero (no correlation), to one (perfect correlation). Apart from the enormous complexities of statistical methods the degree of association itself was always seen to be an insufficient means for determining significance because of such problems as spurious correlations and unknown variables (*ceteris paribus* assumptions).

The second criterion, which in a sense is an extension of the first, concerns the degree to which the generalisation is part of a theory, and the theory, in turn, a part of another theory, and so on. The more a generalisation is imbedded in a theoretical structure, the more confidence we could have about its significance. Hence, the significance of laws and concepts is established in analagous fashions. A law is significant if it is connected to other laws in the sense of being explained by them or if it helps to explain other laws. When laws explain other laws we have *theories*. Some laws in the theory do the explaining as *axioms* and some laws are explained as *theorems*.

Although these criteria have been modified within positivism (as in the Popperian argument about disconfirmation) these two remain the basic positivistic approaches to significance.[7] They, however, do not themselves establish the irrefutable basis of objective significance and if they alone are used as a formula for social science the result may be the analysis of superficial relationships rather than of underlying causes. A realist position offers several amendments to the two criteria which would help make a more accurate representation of the role of concepts, laws and theories.

We stated that theoretically, the law can be used either to explain or predict and that a law pertains only to the subject-matter it addresses. If the subject-matter changes, then the law is still true, but no longer appropriate. It turns out, however, that when a law 'no longer applies' the positivist criteria may not help us decide if it is because the context has changed or if it is that the law itself was incomplete. Therefore, we cannot be sure that the law will successfully predict.[8] In fact the explanations and predictions of the laws of physics work only when the facts they subsume are isolated (that is, when they form a closed system) either naturally, as in the solar system, or under laboratory conditions. Otherwise they will never work out exactly. Laws, then, are really about *potentials* or *tendencies* which the phenomena and their relationships possess. Imbedding laws in theories helps us understand more about the context which laws assume but does not itself remove the supposition that the facts addressed would operate in a closed or controlled environment.

Despite the limitations of our knowledge, natural events continue to happen and to be caused. If we need to know all of the significant factors as we would like to, our laws and theories would be about causally necessary relationships. Recognising that *necessity* and *causality* are part of nature and are what we are after in laws and theories goes farther than most positivists are willing, for they do not consider necessity or causality observable or verifiable conditions but rather beliefs that some need in order to feel confident about laws. This view comes from their definition of the empirical as only the immediately observable or sensible reality; from their neglect of the way we actually use laws and theories; and from their neglect of the fact that natural events will continue to have causes even when man is long gone. A willingness to recognise the existence of necessity and causality and of more or less hidden structures, facts and relationships (some of which could become accessible to our 'senses' as science progresses, for example, as have the concepts germs, viruses and atoms) creates a shift in our conception of laws and theories from one which places logical primacy on the visible and sensible structures and relationships to one which places equal weight on the less visible and less apparent phenomena which influence and control the more visible ones. These realist amendments make us more willing to experiment with theories and radically different interpretations. They show how formulations such as Lévi-Strauss's, Freud's and Marx's are excellent candidates for social science theories if their generalisations account for the facts. But the realist critique and amendments do *not* solve the problems of significance and verification. Rather they make us more realistic about them. The existence of underlying structures still needs to be substantiated by observing the behaviour of the phenomena they are supposed to control. And the regularity of this behaviour is still to be captured in generalisations or laws which will require the same statistical procedures for their partial substantiation or 'verification'. But no longer can the responsibility for verification and significance be placed solely with observed correlations. Rather we must recognise the essential role that is played by critical judgment informed by theory.

Subjectivity, concepts and symbols

Most social scientists and philosophers of science, whether positivists or realists, believe that there is no reason in principle why the study of human behaviour cannot be scientific. In fact, there has been no irrefutable reason offered to think otherwise.[9] Yet there are extremely complex issues to the study of human beings which do not arise in the physical sciences. Even though social science studies the objective facts of nature and objective facts created by people (that is, social artifacts) these are imbued with subjective imports that affect how humans react to them and which in turn affect future human actions. How the subjective is conceived of and handled by a social science is then a critical issue.

From the standpoint of the physical sciences, there is a difference in the objectivity and 'verifiability' of the scientific meaning of mass, as opposed to one's feelings about mass, or between the statements 'It is raining', and 'I am happy'. The difference is inevitably expressed in the distinction between the objective and subjective, between public and private, between physical and mental. Concepts which refer to connotations or directly to mental states (that is, I am happy) are called mental terms and have a different meaning than the meanings$_{1 \text{ and } 2}$ in science. We will call this different meaning, meaning$_3$.[10]

My connotation or subjective meaning$_3$ of the scientific term mass may be different from yours. Similarly, when I say I am happy, my meaning$_3$ of the term may differ slightly or considerably from yours and from a dictionary definition. Whereas our meanings$_3$ of a physical science term may be different, as we noted before, these differences do not seem to interfere with the understandings and evaluations of the physical science definitions or meanings$_{1 \text{ and } 2}$. However, when our primary interest is to understand subjective states and facts we must then consider directly what is the meaning$_3$ of terms or concepts.

To know if a person who says he is happy is in fact happy, we usually compare his use of the term to a standard, clearly defined public meaning. Even if the person does not say he is happy, we may believe him to be, because we know states of happiness to exist and to correspond to certain patterns of behaviour. To know if someone is in such a state may require that the person manifest observable behaviour patterns which conform to a public definition of the concept. Looking for such evidence is what we usually do when we try to find out if someone is 'really' happy' or if he is faking. But the kind of evidence that is needed in everyday conversation and decisions is much looser than the kinds that would be required by scientific descriptions of mental states. Attaining this precision then requires that the meaning$_3$ correspond in precision to a meaning$_1$. From a positivist view this precision can be attained only by describing the mental in terms of physical states (that is, the observable conditions). This means our subjective and intentional meanings can be made scientific by reinterpreting or translating them in terms of precisely defined observable (including verbal) behaviour.

For a person to be happy, according to this view, requires that the person exhibit certain observable biological, psychological or sociological characteristics accompanied perhaps by his own testimony. Eventually such objective sets of facts constituting the meaning$_1$ of a mental state become a 'translation' of the private meaning$_3$ of such states into public concepts and actions. Such associations and translations of private meanings into public ones may apply to a single individual or to groups. Associating and translating meanings$_3$ into clear meanings$_1$ may allow us to find if these meanings are significantly associated with other objective facts and thus enable us to explain and predict subjective states.

A strict positivist interpretation would require that all mental states or concepts be interpreted into physical states. This is because of the primacy positivism accords to physically observable states as the only ones qualifying as real or empirical. Hence the translation of mental into physical becomes a way of instantiating the mental and of eliminating it.

This approach to the subjective takes the mind–body parallelism, as it is often referred to, and makes it unequal, placing the mental in a negative light and leaving the social sciences in the unhappy position of finding much of their own subject-matter, that is, the study of attitudes, values, beliefs, feelings etc., on an unequal footing with the subject-matter of the physical world.[11] The subjective, until it is translated, is scientifically unsubstantiated and suspect.

A realist interpretation would hold that mental states exist even if their translations into physical states is not accomplished. Therefore, not all mental terms need to be translated for them to be instantiated or for them to have clear meanings. Those translations that we do make in our day-to-day dealings with people and in social science can be thought of as disclosing correspondences between the mental and the physical, or as a way of specifying the effect of the mental. For a realist there need be only some translations or correspondences of the mental with the physical to test statements about the mental. From this view it would be expected that theories of human behaviour would normally include mental terms or concepts that were clearly defined (in terms of other mental terms) but which themselves were not translated in physical states or terms.

Recognising the existence of the mental does not, however, minimise the differences between the mental and physical or the subjective and objective, nor does it lessen the need to establish links between the two. What it tends to do is to place the mental in a more positive light. Rather than rush to translate the subjective into the objective it allows us to consider the subjective directly and to explore its own structural relationships. Moreover, there is no need to reduce the causes of the subjective to the objective. The mental, for instance, does not have to be thought of as resulting or emerging from neurological processes. We can consider it to be neither a product of physical process nor an entity of any kind. Rather, according to Langer, we can think of mental phenomenon as an aspect of the occurrence of neural impulses.[12]

That is, consciousness is feeling one's own vital activities and processes. In Langer's terms:

> If one conceives the phenomenon of being felt as a phase of vital process, in which the living tissue (probably the nerve or the neuronal assembly) feels its [own] activity, the problem . . . of how nerve impulse can be 'converted' into thought and thought into nerve impulse . . . becomes a different sort of problem. The question is not of how a physical process can be transformed into something non-physical in a physical system, but how the phase of being felt is attained, and how the process may pass into unfelt phases again, and . . . how an organic process in 'physical phase' may induce others which are unfelt. Such problems, even if far from solved, are at least coherent with the rest of biological [and scientific] inquiry and logically capable of solution.[13]

Moreover, such a view of the mental does not violate the tenets of empiricism. It is in fact completely consonant with the major empiricist position that knowledge is based on the senses.

Among the benefits that come from viewing 'mental phenomena not as a product of neural impulses, but as an aspect of their occurrence'[14] are clearer conceptions of the meanings of subjective and objective. By '"subjective" [could be meant] whatever is felt as action [as having an autogenic origin within] and by "objective" whatever is felt as impact.'[15] There are no classes of things which are solely objective or subjective. Rather, this view leads to the more realistic situation that 'any felt process may be subjective at one time and objective at another, and contain shifting elements of both kinds all the time'.[16]

Conceiving the mental as an aspect of neurological and other vital processes does not solve or dissolve the mind–body problem. Scientifically, to understand the mental still means we must express it in terms of scientifically significant meanings. What this view may do is to postpone the search for translations and correspondences long enough so that we may view the mental in a more positive and fuller perspective, to allow its own patterns to develop so that we can see what it is we are going to analyse. This means exploring the ways in which mental states have been expressed in other modes of thought which are more subjective than the sciences in their analysis and which symbolise feelings and emotions. By symbolise, we mean the use of vehicles such as words, thoughts, shapes and mental images to stand for either subjective or objective things that are not immediately present. This view of the subject directs our attention also to the important and often overlooked consideration that science itself is a form of symbolic expression, one which specialises in symbolising facts (or feelings as impacts) and their relationships. It also allows us to see more clearly that each mode of thought, as a pattern of symbolic expressions, specialises in portraying ranges of the subjective and the objective (the autogenic and impact) in different ways and at different levels of abstraction.

For example, dream images and art symbolise autogenic actions, but in different ways. Dreams are spontaneous, unconscious symbols of autogenic actions. Dreams may be triggered by impacts such as the heat and light of the room in which one sleeps, or by the digestive processes. These kinds of impacts, and the more distant ones, such as the experiences of early childhood, may be incorporated in dream images, but the primary direction of the dream symbol is inward, to the subjective. Images of objective facts and relationships are distorted in dreams, and dreams contain many images which do not seem to represent any specific impact at all. Dream images in general serve as symbols of autogenic feelings which may never surface in our own consciousness. The images are selected and shaped by emotions which often seem to inhere in them. We may not know what an image represents in terms of things or objects, but it may create definite emotional sensations. When a dream image incorporates recognisable, though distorted, elements of reality, it does so to inform us of the subjective.

Unlike dream images, artistic symbols are the result of disciplined, conscious and heightened efforts at abstraction. And although art often represents reality, its primary function is to symbolise feelings. Art, in Langer's terms, 'is the creation of forms symbolic of human feelings'.[17] Works of art, like dream images, often represent actual objects and their relationships. Such is the case in the portrait, the landscape painting, and in the sculpture of a human form. But the primary purpose of these factual details in the work of art is to use familiar representations to draw and hold attention, to fix the gaze, so that we might see the work of art in its most essential function of symbolising feeling. Although a portrait of a king may be the only record we have of how that king looked, as a work of art the portrait would present an image of such feelings as majesty, baseness, power or frailty. The king's features, which in a portrait are never entirely accurately represented, serve as vehicles for such symbols (see Figures 2.1 and 2.2). Sometimes the representation of physical objects and relationships in a work of art may prevent the observer from going beyond the objects to the symbols of feeling. To avoid such situations, the elements of art may be deliberately non-representational as in abstract art.

It is impossible to eliminate the representational elements in art forms like literature and poetry which use words. But representation can be eliminated in music, for musical notes are essentially non-representational. Some sequence of notes may correspond to natural sounds and, on occasions, parts of pieces may imitate such sounds as the cuckoo or the roar of a cannon. The overwhelming structures of music, however, are patterns of sounds which do not represent sounds that are heard in the real world. They have no specific objective referents. Yet, clearly, music symbolises. It refers to something, and that is feeling.

In general, art symbolises feelings, and different art forms specialise in different kinds of feelings and in different realms of life. According to Langer, the plastic arts create an illusion of space or a virtual space, and each one does it in a different way. Architecture creates a virtual place; painting, a virtual scene; and sculpture, a kinetic space. The other arts symbolise other realms of life. Music, for example, symbolises time, and poetry creates a virtual life.[18] The symbols of art are abstract and intentional. Those of dreams are not. The paintings, music, poems and plays are not the facts or the feelings they symbolise. Destroying works of art does not destroy what they represent. The dream, too, is an abstraction but of a much lower order and while we dream we are unaware that dreams are symbols. When we awake, we may even be confused about whether we dreamt something or whether it actually happened.

Man has not always recognised that dreams and art are symbolic forms. Dreams and aesthetic creations have often been conceived of as sharing, participating in and affecting what they represent. A close and often undifferentiated relationship of symbol and referent is characteristic of mythical and magical thought, and historically there is a close association between dreams, art and myth. In myth and

FIGURE 2.2   *Portrait of Philip IV by Velázquez (1631–5)*

FIGURE 2.1   *Portrait of Philip IV by Velázquez (1623)*

The first portrait shows the king to be a rather plain man. The second portrays him with an inner glow as a vision of a king.

magic, symbols refer both to the objective and the subjective and, like dream images, the symbols seem, to the practitioners, to share the properties of what they represent. Hence, in the mythical–magical mode, symbols are often confused with their referents and substituted for them. A circular village, for example, could be seen as a microcosm of the cosmos. This form of thought may be more prevalent in primitive cultures because their limited technology makes their relationship to the world much closer. Hence they tend to see nature with a heightened emotional intensity, making it appear as though objects possess feelings and volitions.

Among small groups of people living close together in the same environment, such projections may be shared by the members of the community and, through such projections, elements of the world may appear to be animated with spirits and forces, and infused with life. The symbols that are selected by the group to represent the objects of their environment may be ones the group feels naturally express both the objective and the subjective qualities of these objects. The circle, for instance, may be the symbol of the cosmos. It may represent the apparent roundness of the heavens and be symbolic of perfection and the eternal. So close may the symbols appear to resemble their referents that they may themselves be seen to possess the qualities of what is represented. In such cases, the creation, manipulation or destruction of symbols may be expected to affect the referents and other objects. Creating circular settlements is expected to recreate the cosmos and its heavenly forces, words are used to cast spells, and sticking pins in dolls is believed to result in injury to the person the doll represents. Mythical and magical systems are very often enormously complex but the mode of thought never frees itself from the close association of symbols with facts and feelings and of the conflation of symbols with them. The mythical–magical mode conceptually separates subjective and objective, but only partially and haltingly.

The extreme undifferentiation of the subjective–objective is the mark of early childhood and childlike thought. Infants barely distinguish between feelings as impact and feelings as autogenic action. The separation at first occurs gradually, makes room for, and then is accelerated by, the use of symbols. When the child first uses symbols such as language, many of the words seem to be confused with what they represent. The young child may not understand that one can be talking about someone or something that is not immediately physically present. When a person's name is mentioned in a conversation, the child may look to see if the person is present. The child so closely associates the symbol with what it represents that he may expect the referent of the symbol to be present when the symbol is used. Similarly, the child may believe that simply using words can physically affect an inanimate object. The child may tell the sun to rise or the wind to stop. But the normal child grows out of this phase into an adult with a practical view. One of the characteristics of adults is their use of natural–ordinary language for practical ends.

Natural language is the most characteristic symbolic form of the practical mode. In the normal practical use of language, the symbols of language separate the subjective from the objective, but not as extensively as do the symbolic forms of science and art, for the terms of natural language, no matter how clearly we wish to define them, and how specific their referents are, refer to real relationships and retain strong emotional resonances. If we were to ignore such connotations as completely as possible, such terms would no longer be part of a natural language but would constitute terms in the artificial language of science. On the other hand, emphasising the connotations and the subjective referents in order to capture the variability and fluidity of feelings, leads from natural language to the language of poetry and literature. The combination of connotation and denotation in natural language leads to ambiguity, but the growth and vitality of language depends on such ambiguity. It is the source of metaphor.

It is in science and art that we attain high, sustained abstractions which sharply distinguish between subjective and objective. Yet, even in science we find that the subjective plays an essential role. As we have said, it aids in the context of discovery; but more than this, it enters into the realm which, at first glance, seems especially immune from the subjective, for it forms the basis of logical inferences which are the foundation of scientific reasoning. It is difficult to believe that logical certainty is found in reality and that our sense of it comes from experiencing it. Rather, it seems that logic involves a pattern of autogenic feelings which are so specific and universally experienced that upon reflection we find it difficult to call it feeling at all.[19]

Examples of the use of symbols in logic, science, myth and magic, art, the child's view, and the practical view, demonstrate, as we shall discuss further in subsequent chapters, how conceptualisation, in each mode of thought, to varying degrees separates and emphasises the subjective and the objective. The points we are interested in emphasising here are that these modes can be explored if we are not in a hurry to interpret every isolated expression of the mental into objective terms. The view of the subjective as a phase of vital action, the reinterpretation of the objective and subjective as impact and autogenic actions, allows us to spend more time on the subjective. By considering the patterns and structures of symbols in modes of thought which represent the subjective, science can at least obtain an image of some of the patterns and interrelations involved in feeling.

But as we noted before, this view of the subjective and the investigation of its symbolic forms only postpones the problem of scientific interpretation. Instead of translating isolated mental states and terms into scientific states and meanings, we are now confronted with interpreting the meaning of complex subjective symbolic forms and expressions in different modes of thought. This has proven to be as formidable a task as the interpretation of isolated subjective states and it raises again the schism between the objective and the subjective. The problem again revolves around meanings. At this level it can be thought of as an asymmetry between the way a symbol refers to the subjective and to the objective. This asymmetry characterises and separates those modes of thought specialising in one or the other. In both natural languages like English and French and in artificial languages like science, statements can be made in propositional form to express facts. These statements follow clearly defined rules of language. They contain a vocabulary or a set of clearly defined terms and adhere to rules of syntax and grammar which determine proper and improper combinations of terms.

Modes of thought symbolising feeling, however, although they may use the terms and rules of language, as in poetry, literature and drama, do not, in their artistic import, contain a set of clearly artistic terms which serve as elements of the work of art, nor do they adhere strictly to explicit rules defining permissible statements. The differences between the two forms of expression can be indicated, for instance, by the fact that the scientific concept mass has the same objective meaning in many statements in physics, such as $F = MA$ and $FT = MV$, where F is force, M is mass, A is acceleration, T is time, and V is velocity. Similarly, in science there are by definition 5280 feet in a mile and this number of feet can be substituted for the term mile without loss of objective meaning. I walked a mile or I walked 5280 feet have the same referent. But when our interests are in connotation and the subjective, as they are in the arts, substituting words with the same objective referents may alter the meaning and sense of the work of art. We cannot substitute 5280 feet for a mile in

> A mile behind is Gloucester town
> Where the fishing fleets put in,
> A mile ahead the land dips down
> And the woods and farms begin.
> Here, where the moors stretch free
> In the high blue afternoon,
> Are the marching sun and talking sea,
> And the racing winds that wheel and flee
> On the flying heels of June.[20]

The different way in which symbols refer is seen even more plainly in the arts which do not employ languages. A painting, for instance, is composed of colours, forms and patterns. But each of these elements does not have a standard meaning. Their meaning is derived from the context of the piece of art, and in great works of art, the elements may have a multitude of meanings. A dash of red or an upward curving line in one painting may contribute several things to a painting, while the same or a similar dash and line would contribute quite different things in another. Moreover, it is extremely difficult in works of art to decide what are the elementary building blocks of meaning. Is it a colour, is it a line, or part of a line, or is it the relationship of these elements to their surroundings? Since these elements would have different meanings in different contexts, there is no set of isolatable terms in art that can be dealt with as are the terms in languages which represent facts.

These asymmetries in symbolising facts and feelings have been referred to by Langer as the difference between *representation* and *presentation*.[21] Science represents objective facts and relationships. The meanings of its terms are matters of convention. On the other hand, most of the symbols and elements of art are not conventionally or unambiguously defined. Nevertheless, art and dreams symbolise feelings and the meanings of the symbols seem to inhere in the symbols themselves. The symbols of the subjective seem to present their meaning while the symbols of the objective represent theirs. The meaning of an equation in physics such as $F = MA$ is not intelligible unless the meanings of the symbols are made known, as when we say that F is force, M is mass, and A is acceleration. The symbols, as conventions, do not physically resemble what they represent. We could alter the equation's symbols to $P = RS$ without any loss of meaning if we understand that $P = F$, $R = M$ and $S = A$. But a dream or a painting can present feelings without the need of an exegesis, and altering its details would change or distort its import.

Such differences in symbolic functions may make it as difficult for social science to understand and explain art and dream images as it is for social science to find corresponding physical states for mental states. While the patterns of feeling, as symbolised in other modes of thought, may be as immune from scientific penetration as the mental facts they symbolise, exploring the expressions of feelings in other modes of thought at least could reveal the variety and pattern which expressions of feeling can attain and could give us a clearer idea of what needs to be translated.

The manner in which symbols are used to conceptualise, and hence, to conceptually separate subjective from objective referents in the different modes of thought, affects the way in which space and substance are conceptually separated and conceived in these modes. There are meanings of space which are abstract and relatively objective or subjective, and there are meanings of space which combine both objectivity and subjectivity and which do not conceptually separate space from substance to a great degree. The highly abstract and intense conceptual separation of the objective, the subjective, of space, and of substance, and the difficulty in affecting their recombination in the sophisticated–fragmented modes has created tensions and conflicts in meaning which are geographically manifested. In addition to mythical and sacred places, there are places created for science and technology, those created for art. There are places that are aesthetically designed and those that are functional. But the subjective and objective are never completely separate, and sometimes in Western societies they are successfully merged. Places can appear, to many, to be aesthetic and functional. And even when such mergings do not succeed, one mode may be the inspiration of another. Feelings about science and scientific concepts of space have inspired artistic forms to symbolise them, and the new forms in turn have been examined by science for their functional utility.

The subjective and objective appear to have been more thoroughly fused in primitive societies which have not developed compartmentalised symbolic forms like art and science. In these simpler societies there may be a greater consensus about the subjective and objective meanings of symbols. A structure or place in the landscape can be appreciated aesthetically, functionally and mythically. But in modern society, while we may personally combine several modes in our appraisal of a single object, our conceptions of things tend to be segmented, and our landscapes reflect this by having places for specific modes. Rarely are appraisals resulting from a combination shared throughout the culture.

The various conceptions of space and their interrelationships can be understood in the context of the conceptual surface, in the way in which space, substance, subjective and objective are conceptually separated and recombined. The mechanisms which extend the axes and surface and which create the different meanings of space are the forms of symbolisation. Having discussed some of these forms as they pertain to the conceptual separation of subjective and objective, we will now consider how these forms extend the space–substance axis and how space is conceived of in different modes of thought.

3
Social Science and Objective Meanings of Space

Legacy from physical science

Scientific thought conceptually separates, analyses and attempts to recombine facts and relationships. This is true of scientific thought about space. A scientific conception of space depends on how space and its properties are thought to be interrelated with the physical categories of substance and time in the context of laws and theories. These interrelationships, although more or less agreed upon by physics for activities at the geographic scale, remain controversial when applied to the extremes of the physical universe, the astronomic and the sub-atomic. The core of the controversy in physics is over the distinctions between absolute and relational space and in action by contact and action at-a-distance. We have already mentioned that the social science conception was influenced by action, by contact and the law of the conservation of energy. It also has been indirectly affected by the controversies over relative and absolute space. Both of these issues therefore affect the social science view of the space–substance axis. We should briefly consider these issues before examining in detail the social science conception of space. We will begin with an overview of absolute and relative or relational space.

In very general terms, absolute space means that the continuum called space is immune to influence, that its structure is rigid and cannot be changed by matter or energy, and its description in geometric terms is independent of one's viewpoint or frame of reference. In a stronger sense, absolute space also means that space can exert physical effects. Relational space, on the contrary, assumes that space can be acted upon, that its properties and descriptions are dependent on the distribution of mass and energy and that, by itself, space therefore would not exert physical effects.[1]

With the advent of relativity, it appeared that the relational view had triumphed over the absolute view, for the nature and form of space was to be everywhere dependent on the existence and distribution of matter. This conclusion seems to have been too hasty. As Einstein suggested, there has been no clear victory.[2] The replacement of the absolute with the relational is still an uncompleted programme. The dependence of space on matter and energy has found only limited expression in the General Theory of Relativity. While space is relational in the sense of its association with time in a space–time system, there seems to be as yet no way of completely eliminating the concept of absolute empty space from the equations of relativity. For instance, it is possible to speak of the shape of space when matter is absent, and it is necessary to assume boundary conditions about space to solve field equations. 'Although matter provides the *epistemological* basis for the metrical field, the fact must *not* be held to confer *ontological* primacy on matter over the field: matter is merely part of the field rather than its source.'[3] It should be noted that the conception of space is even less settled in the theories of sub-atomic physics. Here the same distinctions between absolute and relational space may not even apply.

Controversy about the relationship between space and substance is at the core of the distinction between action by contact and action at-a-distance and, as we noted in Chapter 1, action by contact is at the core of the social science and practical views of space. Action by contact is self-explanatory in cases where the cause and effect are actually spatially contiguous as when one billiard ball collides with another, imparting its momentum to the seond. If two interacting bodies, however, are not physically contiguous but separated from one another, action by contact assumes the existence of some knowable medium connecting the two and through whose 'active participation' the effect is communicated. When two people are speaking to each other, they are physically separated and it may appear, superficially, as though there is action at-a-distance. But the air masses, by transmitting energy through their compression and expansion, are the connecting participating media which make normal conversation an example of action by contact. Similarly, communication by a telephone, radio or television are all examples of communication by contact. Action by contact is expected from the law of the *conservation of energy*. This means that if energy is transmitted from one point and takes a finite time to reach another point away from the first, the energy should be located somewhere in space between the two points between the time of transmission and the time of reception. Hence, the space must be characterised by energy.[4]

Action at-a-distance is taken as a serious issue when we do not have clear, unequivocal knowledge of the intervening causal links and the participating medium.[5] It arises, for instance, in the interpretation of gravitational attraction. In this case, it can appear as though the force of gravity could occur without the participation of a medium, that it could occur through empty space. Since gravitational attraction decreases as the square of the distance between bodies increases, where the force between two masses m_i and m_j is given by the equation

$$F = K \frac{m_i m_j}{d_{i.j}^2}$$

action at-a-distance interpretation would mean that a property of this empty space, namely distance, by itself, has an effect. Modern field theory has changed the conception of medium and action and offers an action by contact interpretation of gravitational attraction and other forms of attraction which were previously seen as action at-a-distance. As with the case of relative and absolute space, the issue is not, yet, unequivocally decided. This indecision though has not prevented a physical science consensus about a conception of space appropriate to terrestrial–geographic scale events.

Action by contact, the conservation of energy and the relational space coalesce in the physical science description of terrestrial space. From the physical science viewpoint, space, for terrestrial or geographic events, is appropriately defined as three-dimensional Euclidean in which action occurs by contact. Action by contact is expected because human interactions are based on the transmission of energy at a finite speed through well-established media and channels. In this regard, distance in and of itself, is not expected to affect human behaviour. Moreover, the general Euclidean structure of space is itself immune from the effects of human behaviour because these activities, according to physics, are of insignificant mass and energy to affect the geometry of space. The social sciences therefore receive a view of space, the characteristics of which are beyond their purview to explain. Receiving this view, and expecting this conception of space to be significantly related to the substances of human behaviour in the same way that it is related to the substances of physical sciences has created problems at the levels of description and explanation. These problems have made it difficult to conceptually recombine the end points of the space–substance axis.

The idea of space that the social sciences receive from physics is clearly defined and accurately measured. In fact, it is far more clear and accurate than the social science facts that are to populate it. Uncertainty about the spatial properties of social science facts is often due to uncertainty about their meanings. Clearer meanings would make their spatial properties clearer. We would know more precisely where a village begins and ends if we were more precise about what we mean by a village. But much of social science is concerned with the subjective, and here, at best, there is only an indirect link with physical space. Objective knowledge of subjective conditions depends on there being a correspondence of such conditions or states with physical states. Hence, the geography of the subjective states would be the geography of the objects and events which manifest them. For instance, we seem uneasy about saying that a belief actually is locatable in space in the sense that it has material being with position and shape. But we do not feel uncomfortable about calling a person or group of people religious and of making maps of religious adherence. In such cases, the location of the mental states (religious attitudes) is in terms of the people and artifacts which we associate with such states.

The ends of the space–substance axis in the social sciences are also uncertainly linked in the area of significance. We know that the location of things affects their behaviour; that moving them from one place to another makes a difference; that changing the spatial configuration or pattern of social arrangements alters their functions; and that physically separating two interacting objects will affect the degree of interaction. But there are difficulties in expressing these relationships in a social science so that they conform to the principles of action by contact. The laws of social science cannot violate the laws of physics, and physics contends that the activities of human behaviour do not affect the properties of physical space. How then are the social sciences to express the fact that people conceive of the nature of space differently according to different viewpoints and goals? And how are they to express the fact that the same spatial relations may be evaluated differently in different modes of thought? These are important matters: what people believe to be the case influences their behaviour. The answers to these questions, and to that of the general conception of space in social science, are to be found in the way physical space is connected to social science facts or substances, in social science explanations or laws. We will refer to this connection as the *relational concept of space*.[6] Definitions and laws in science are the means by which space and substance are conceptually recombined. Since there are as yet no powerful social science laws or explanations to use as illustrations, the analysis will have to rely on a very general model (derived from the realist view of the science discussed in Chapter 2) of what social science explanations might look like, given the present state of the disciplines and their aspirations to become like the physical sciences. The model will concentrate on the level of laws and will be amended as we discuss the more subjective views of space.

The model and the relational concept of space

Spatial terms, substance terms

Reflecting the tripartite division of time, space and thing, the empirical terms of the sciences, social sciences and even normal language can be partially divided into three categories – temporal concepts, spatial concepts, and the vast remainder of non-temporal, non-spatial ones, which we shall refer to as substance terms or concepts. Temporal terms refer to such units of measure as minutes and hours, while spatial terms refer to such concepts as points, lines and shapes, which, for terrestrial events, are scaled in such commensurable units as metres or miles. By *substance* terms are meant all other non-logical terms referring to the objects, powers or facts of the empirical world – the fields, roads, institutions, values and attitudes which form the subject-matter of the social sciences, and the masses and energy of physics.[7]

Although empirical concepts are conceptually separable into these categories, the three are drawn together in the identification and individuation of facts. This interconnection is based on the fact that substances are in space and time and that space is not instantiated independently of substance. Nevertheless, a fact can be described differently in each of these categories. A fact, such as Harry's ten-year-old house, is describable in substance terms as a house, in temporal terms as ten years old, and in spatial terms as a cube, located at coordinates x and y. Science discusses facts in terms of concepts, and it is important to our purposes to explore how the factual connection of substance and space especially are reflected on the conceptual level through scientific concepts and generalisations. We will concentrate on the link between substance terms and spatial terms, and begin by emphasising the substance side of the relationship.

As was noted, the meaning of concepts can range from the simple procedure of ostensive definition to the more elaborate operational definitions. Like any other terms in science, the definitions of substance concepts are made with an eye towards significance. Substance concepts abstract those properties of things which it is hoped will lead to generalisations. We can divide substance concepts into two categories, those that do not explicitly abstract any spatial properties of the substance and mention them in their definition, and those that do. Those in the second category, of course, cannot include specific places because concepts are abstractions about classes of facts. Rather they can include general spatial relations such as lengths and shapes.

An example of the first kind is the concept 'nation state'. We know that nation states must be located and occupy space, but there is nothing in the definition or meaning of the concept which specifies what shapes or sizes the states are to be. There are large states, small states and prorupted states. It will be remembered that a scientific concept has to be sufficiently well defined in order for us to know if instances of it occur. Even though concepts may not mention a spatial characteristic, the instances nevertheless have spatial properties.

In concepts without explicit spatial terms in their definitions, the spatial properties of the instances are unforeseeable from the meanings of the terms alone. Each instance may have entirely different spatial properties. Hence, from the meaning of the term nation, we know nothing about the actual shapes of nations, their locations and sizes. Yet, whenever there is a nation, this nation is in a place and has a shape. To state that an instance of a substance term occurs is to imply that it occurs somewhere and at sometime. (There is, of course, the third dimension which is usually omitted in social science.)

It may be difficult precisely to locate and determine the spatial properties of some facts. Such problems may be due to the vagueness of a concept's meaning. The vaguer the definition, the more difficult it is to know an instance of it, and hence, the more difficult it is to know where the thing occurs. If we are not certain what we mean by a city, we will be uncertain about where cities are, or what exactly are their extensions, where they begin and end in space. Vagueness about the meaning of things leads to vagueness about their spatial properties.

In addition to vague concepts, the social sciences have many open-ended concepts. A city, for instance, may be defined clearly as a market centre which supplies goods and services to a hinterland and in which there is role specialisation and division of labour and high population density, etc. . . . . The 'etc. . . . .' points to the fact that the meaning is open-ended and so too will be the spatial properties of the city on the ground.[8]

Some substance concepts in the social sciences and in natural language include spatial terms in their meanings. These form the second category of substance terms. The term 'dome' is defined as a rounded vault. Hence, whenever we find an instance of a dome, by definition, we find an instance of the spatial concept 'roundness'. By definition, such concepts contain more spatial information in their meanings than do the first category. Instances of all concepts are spatial, that is, they have location and extension, but those of the second category may be more regularly shaped. Whatever else domes may be, they will be rounded, which is not the case for nation states. The concepts or terms in the second category, like the first, do not specify where in space and time the instances occur, nor do they stipulate completely the spatial properties of the facts they subsume. Even though all domes are rounded, they may be large and small, thick and thin. If all of the spatial properties were specified, the concepts together would provide an exhaustive description of a single instance.

Since a fact, expressed as an instance of a substance term, would have spatial properties, the same fact can be expressed as an instance of one or more spatial terms. Thus, a cube-shaped house may be expressed on the one hand as an instance of a substance like, 'This is a house', and as an instance of a physical spatial term like, 'This is a cube'. Because, for terrestrial events, space is not empty and actions occur by contact, we can also assume that every part of space is also a 'part' of one or more substances. To put it in terms of concepts, each and every instance of a physical geometric term must be connected to one or more instances of substance terms. This is the way the end points of the space–substance axis are conceptually related in terms of identification and individuation. When we say '*that* triangle over *there*', we will be able to find *there* one or more substances of which this triangle is made or with which it overlaps. It may be a triangle made of light waves in the sky or a triangle made on paper and crayoned a red colour. The light, the colour and the paper are the substances coincident with the triangle. What makes the relational concept in social science is the way in which this symmetrical relationship between instances of substance terms and instances of physical spatial terms become significant. By significant, we mean the way in which spatial terms are incorporated into hypotheses and laws of human behaviour – the way in which laws conceptually recombine the end points of the space–substance axis. Understanding this also clarifies why some substance concepts include physical spatial terms in their meanings and others do not.

Laws

Paralleling the distinction between spatial terms and substance terms are two sets of laws that use these terms – the laws of *physical geometry*, and the non-geometric laws of science which we shall call *substance laws*. These two sets of laws explain facts differently, and this difference points out an asymmetry in the way facts can be subsumed within substance or geometric vocabularies.

Geometric laws   Facts, as instances of spatial or geometric terms, can be explained by the laws of physical geometry. The terms of physical geometry are not only clearly defined but they are also significant within the laws of physical geometry. These laws form a relatively complete, closed, deductive axiomatic system. Their only empirical concepts are the physical geometric. These laws are static. They do not explain processes. For instance, the Pythagorean Theorem is about a static relationship. If we know the lengths of the sides of a right-angled triangle, it allows us to determine (or predict) the length of its hypotenuse. Similarly, if two interior angles in a triangle are measured, the third becomes known without the need of measurement because the sum of the three interior angles must equal 180°. From the relational concept we assume that every instance of a physical spatial triangle would also have to be an instance of one or more substances. But the laws of the right-angled triangle, or any other Euclidean geometric relationship, apply independently of the substances which constitute or embody the geometric relationship. It would make no difference whether the right-angled triangle were made of steel, of paper or of light. The laws would hold for all the substances, and hence, are formulated independently of the substances. For terrestrial events, then, although space is related to substance, the laws of space operate as though space were unaffected by substances, as though space were empty.

Geometric laws explain and predict the physical facts as instances of physical geometric terms. The explanations are static, immune from process and are removed from the direct concern of the social sciences. Geometry would explain the shape of a building in geometric terms. Its rectangularity would be due to the fact that its angles are 90°. The explanation tells us about the walls of the building only as an instance of angles and rectangles. It tells us nothing about why, in terms of human behaviour, the building is rectangular.

Substance laws and spatial observational instructions (congruent) The social sciences search for answers about spatial relations that are not purely geometric. Social science conceives of facts as instances of substance terms, and the laws in the social sciences, as well as in physics, which would use such concepts are *substance laws*. The rectangular shape, for instance, would be conceived of as the rectangular shape of a house's walls, and these are substance concepts. The occurrence of the rectangular substances would be explained by subsuming them as instances in generalisations of laws about house types and preferences for such types. While substance laws always include substance terms, they may also include spatial terms, like 'rectangular', or like 'distance'. If we are to know the significance of space and its properties in social science, we must determine what spatial information there is in social science explanations. This involves understanding how and why geometric terms are selectively included in substance laws. Spatial information is contained in substance laws in what we shall call their *spatial observational instructions*. Stipulating and following these instructions is the *relational concept of space* in social sciences and outlines the conceptual recombination of space and substance in the context of laws.

Substance laws or hypotheses are conjunctions of substance concepts. Two kinds of substance laws appear to underlie the social sciences: those that do not include explicit mention of physical spatial terms and those that do. The first kind generally have the form of 'if *A* occurs then *B* occurs', wherein the generalisation does not mention where, relative to the antecedent conditions, the consequent conditions will occur. Nor does it mention the shapes, forms, patterns and distances of the facts explained by the law. This is the case in such hypotheses from the social sciences as 'If freedom of the press is denied, fascism will occur', 'If a person has a high IQ, that person will likely do well in the first year of college', 'In a free market, if the price of a good increases, less of it will be demanded'; and in laws from the physical sciences such as those about electricity where 'The energy times the current equals the wattage', and 'The energy divided by the current equals the resistance', and that about freezing water where 'Water, under standard pressure conditions, will turn to ice if its temperature falls below 32° Fahrenheit'.

Although such generalisations say nothing explicit about the relative locations and shapes of the antecedent and consequent conditions, they do include such information implicitly, for if we were to test any of these hypotheses we would expect to find the consequent conditions occurring where the instances of the antecedent conditions occurred. We would expect the resistance and the wattage at a particular point to affect the current at that point; we would expect that wherever standard pressure conditions held and the temperature of water was below 32° Fahrenheit, there would be ice; wherever freedom of the press was denied is where fascism would occur; and where there is a person with a high IQ, there is a good chance of academic success. In such cases we can replace the 'if' and the 'then' by the term 'wherever'. The point is that omission of explicit spatial terms does not mean such hypotheses are without spatial information. In fact, they contain very powerful spatial information. They tell us that the instances of the consequent conditions would occur at the same place that the instances of the antecedent conditions occurred. We refer to such spatial information that a law provides as the *spatial observational instructions* of the law or hypotheses. In general, the spatial observational instruction of laws which do not explicitly mention physical geometric terms or concepts is that the facts subsumed under the consequent conditions of the law (the instances of the antecedent and consequent conditions) are physically spatially *congruent* with one or more of the facts subsumed under the antecedent conditions of the law. These facts occur in, or occupy, the same area. Such laws will be termed *congruent laws* and comprise one of two categories of substance laws based on the spatial information in these laws.

Congruency, of course, does not explain all of the spatial properties of the facts, just as no law explains all of the properties of any fact. They only explain that which is subsumed by the concepts. If, for instance, we were interested in asking questions about why the person with a high IQ is where he is, or why the water that froze is where it is, we are asking questions that go beyond the original generalisation. These questions are no more spatial than the congruent ones which associated IQ with success, or ice with temperature and water: and they, in turn, may be explained by other congruent laws, as when we would explain the location of people by a theory of residential preferences, or the occurrences of some bodies of water by glacial scouring.

Often the hypotheses may not mention anything about the temporal order of the facts, implying that they are to occur simultaneously. If the facts are indeed simultaneous it might appear that for them also to be congruent would violate the assumption that no two objects can occupy the same place at the same time. In the social sciences this problem is avoided because the location of facts often is based on aggregating them within an area and assigning the aggregation to an areal unit such as a political territory. When we say that freedom of the press is denied in this or that country we mean that the presses, which are located within the country's boundaries, are censored. Similarly, fascism occurs within the boundaries. While the presses, the citizenry and the reporters are in different places, they all are located within the territorial bounds of the state. Since the areal unit for which these facts are expressed is the state, these facts are accordant at that level of aggregation. The same kind of accordance is involved in saying that an individual with a disease has several symptoms: each may be in a different area of the body, but they all occur within the same individual.[9]

What appears to be a congruent hypothesis may hold only at one scale of precision. If looked at on a smaller scale, it may require additional spatial information. For instance, when water changes state from a liquid to a solid, it actually occupies a larger area. So the spatial relationships among the facts in the ice, water, temperature, pressure generalisation are more precisely described as overlapping, not congruent, and a complete expression would include this spatial information. Similarly, a more detailed analysis of social science generalisations which involve data that can be disaggregated may change some congruent hypotheses to noncongruent ones. But this would happen only to some. The important point is that congruence does occur in fact and it is a spatial relationship specified in scientific generalisations and laws.

Substance laws and spatial observational instructions (non-congruent or contiguous)　When the antecedent and consequent conditions are not in fact congruent at a certain scale, the hypothesis or law would contain one or more physical spatial terms to so stipulate or direct our observations. These hypotheses form the second category of substance laws which we will term *non-congruent or contiguous laws*. Such laws may be sketched as 'if A occurs then B occurs X distance away from A'. Examples of the non-congruent relationships in social science are the generalisations of proxemics, which discuss the positions of people in personal encounters; the geographic generalisations relating land-use practices to transportation costs and to distances to market; generalisations relating interactions among places to the distances between them; and generalisations relating sizes of central places, towns and cities to their geometric arrangements.[10] In physical science we find them in such laws as Coulomb's law of electrostatics, in which the force F between two charged bodies q_1 and q_2 which are a distance r apart is proportional to

$$\frac{q_1 q_2}{r^2}$$

and in the law of the periodicity of the pendulum where the pendulum's period is equal to

$$2\pi \sqrt{\frac{l}{g}}$$

l being the length of the pendulum and g, the force of gravity.

All of these formulations explicitly include such spatial terms as distance because the events and substances subsumed under the laws or hypotheses are not spatially congruent. Rather, they are separated. And as with the congruent relationships which excluded explicit mention of spatial terms, the second category of laws, by including spatial terms, tells us where, relative to the antecedent conditions, the consequent conditions will occur. These messages are their spatial observational instructions. In general, when the facts are not congruent, the laws usually say so through the explicit inclusion of physical spatial terms, which state where, relative to the antecedent conditions, the consequent conditions will be. When the facts the law explains are spatially congruent, the law omits any explicit mention of space.

While the use of spatial terms signals the non-congruency of the facts to be explained, defining the meanings of these spatial terms, how they are to be included, and to what they refer is the basis of the relational concept of space in social science. Unlike the use of spatial terms in the laws of geometry, the significance of spatial properties like distances and shapes, depends, in substance laws, on the things or substances of which these are shapes and distances. These things are the substance referents of the spatial terms. In substance laws and generalisations, the significant substance referents of the spatial terms must be either mentioned in the generalisations or understood in the context of the generalisations. *Including the significant substance referents of the spatial terms is the relational concept of space in social science.* We add 'significant' substance referents because a shape or distance may be manifested by several substances. The one we are interested in is the one that significantly affects the other variables in the relationship.

Take, for example, the generalisations from proxemics in which people from the same culture usually stand a certain distance apart in normal face-to-face conversations. The significant substance referent of the distance in this case is the air molecules, since they constitute the medium through which the energy is transmitted. The generalis-ation of proxemics may not make this explicit because we know more than the generalisation tells us. If, for example, the speakers were the proper distance apart and there was a one-foot thick pane of glass between them, a conversation would not be likely, nor would we expect it to be, because we know that sound does not travel as well through glass as it does through air. If there were no panes of glass between the speakers but instead the air were heavily foggy, then the distance separating the speakers might be less than if the air were clear. In cases where we have control over the process and know it well, we need not remind ourselves of significant substance referents of the spatial terms. But in most examples of human behaviour we do not have such control or information and here stating the significant substance referents of the spatial terms is necessary in order to generalise about behaviour.

When two or more objects causally linked are not congruent in space, but rather separated, the appropriate interpretation according to action by contact is that the action is continuous through substances in space. Having the objects separated and including spatial terms to so stipulate this does not mean that the objects are separated by empty space. Rather, they are connected by intervening substances or mediums which have lengths. The objects are separated by other substances. Hence, the entire system of interaction is more properly thought of as a spatially contiguous relationship. This is why we refer to the second type of substance generalisation or law as a *contiguous* relationship.

The critical point for the social sciences is that these substances or media through which things are transmitted have a bearing on the transmission. In our proxemics example, all the people who are physically separated are separated by air and the quality of the air could affect conversation. Sound travels at different speeds depending on the temperature and pressure of the air, and our ability to hear the message depends on the other sounds that are in the air. In this regard the effect of distance or space depends on substances. Substances affect the significance of space but they do not affect the description of space. Two people three feet apart will be three feet apart if the distance is of air, or of glass. However, if it is the latter, they will not have a conversation, nor may they be expected to remain three feet apart for long.

The relational concept of space is borne out in our practical assessments of distance. We look at the distance to a place to which we might wish to drive in terms of road miles, not straight-line distance. We consider road condition, traffic, scenery and other substances we might encounter. The interpretation is also borne out in laws of physics. Here we find that though such spatial properties as distance are variables in equations, these properties are not independent agents. The nature of the medium through which energy passes makes a difference. The force of attraction between two electrically charged bodies depends on the distance separating them, *and* on the nature of the substance (or medium) of which this distance is a measure. The time it takes for sound to travel from one point to another is a function of the distance and atmosphere between them. There are, of course, many occasions when the substance referents of physical spatial terms are clear from the context, and need not be mentioned explicitly as a condition of the law. In the law explaining the period of the pendulum, for instance, distance is introduced as a variable in the form of the length of the material that suspends the bob. Although the law often makes no explicit mention of the substance of which this is a length, it is understood that the material will affect the pendulum's periodicity through such factors as differential effects of temperature on the expansion and contraction of the material and that such effects can be included in the law if need be.[11]

In the physical sciences, the effects of the substances of which distance and other spatial properties are connected are well known and may either be included in the context of the generalisation or else be omitted. But in the social sciences, where there is not an abundant knowledge of significant concepts or even full knowledge of the paths and networks of interaction, we cannot afford to omit what are believed to be the significant substance referents of the physical geometric terms in our hypotheses. If they are omitted, as has often been the case, we might be led, as others have been, to the erroneous conclusion that properties of space themselves have an effect on behaviour.[12]

To sum up then, the end points of the space—substance axis would be conceptually recombined in the definitions and laws of social science. The laws of social science would be substance laws and among them we have two types, those that do not explicitly contain physical geometric terms in their concepts, and those that do. Both types nevertheless have spatial information in the sense that they explain and predict where, *relative* to the consequent conditions, the antecedent conditions are or will occur. I emphasise the word *relative* because neither the congruent nor the contiguous laws state where exactly in space and time the instances will be. To do so would be to include specific locations, and laws are not about the specific; they are about the general.

Those laws that omit explicit reference to spatial terms we refer to as congruent substance laws. These laws imply that instances of their consequent conditions have the same locational—extensional characteristics as those of their antecedent ones, at the moment the consequent conditions occur. Although none includes explicit mention of geometric terms, their spatial directions and their ability to explain and predict the geometric properties of the instances of their concepts are clear: for in each case, the locational–extensional properties of the consequent conditions are expected to be congruent with one or more of those of the antecedent ones at the time the consequent ones occur. Furthermore, all of the concepts have significant spatial implications in the law. When the antecedent conditions precede the consequent one in time, the congruence is often suggested by such phrases as the antecedent conditions 'evolve', 'become', or 'change into' the consequent ones.

When the instances of the substance concepts in a law or hypothesis are not congruent, they include physical geometric terms to so stipulate this. The inclusion of explicit geometric terms, such as distance, must be accompanied by the mention of the significant substance referents to satisfy the relational requirement. This means that the facts explained in such laws are not separated by empty space, but rather are connected by intervening substances. Thus, all the facts in a law, though not congruent, are connected in physical space. They are contiguous and form chains or linkages which conform to the principles of action by contact. The spatial extent of the chain is specified by the spatial variables in the substance generalisations.

As with the case of the congruent laws, each concept in contiguous laws, whether geometric or not, is equally important in determining the locational–extensional properties of the instances of the other concepts. The locational–extensional properties of any one of these depend on, or more literally are determined by, spatial properties of the others. The concepts are significantly interrelated to form a connected and extended system in space.

The congruent and contiguous laws then are the two forms of spatial relations in social science generalisations or laws. From the spatial–observational instructions and the relational concept we can determine how a social science with generalisations or laws would conceive of space. In the first place, statements that geographic location has an effect or influence on human behaviour, or analogously, questions which emphasise the location of an event like, 'Why does this occur here?' rather than asking simply, 'Why this occurred', cannot meaningfully suggest that it is location itself that is important or that the answer to the question of *where* or *here*? will be different from the answer to the question *occur*? To explain why something occurs is to explain why it occurs where it does. Similarly, such locutions as 'Location affects behaviour', 'Distance affects interaction', or 'Form affects function', would have to be specific regarding the substances manifesting these properties for the statements to make sense in the social science. Knowing spatial properties such as form or pattern without knowing the substance referents will not reveal processes. Similar processes may generate the same form or different forms and the same form may be generated by different processes.[13]

All substance laws would explain or contain spatial information. If we define the spatial power of laws as the ability precisely to explain and predict physical spatial properties of the facts they subsume, we cannot distinguish between congruent and contiguous laws in terms of their spatial powers. Both sets are equally spatially powerful.

We mention these factors because we so often hear social scientists speak about the effects of space and its properties as though they could be isolated within a social science context and given a determinate significance. The closest that we can come to saying that a spatial property is, in and of itself, significant, and still be in conformity with the relational concept, is to find a condition in which the effects of a spatial property such as a distance or a shape will not vary at all for all substances of which it is a length or a shape. Such a condition would mean for example that a ten-foot length of anything and everything will have the same effect on behaviour or that a circle, regardless of what it is made of, will always produce the same feelings and beliefs. If such cases were to exist in fact (and we will explore the subjective realm to see if shapes and feelings are associated) we may then, to all intents and purposes, speak as if it is the distance or shape itself that affects behaviour. But there is as yet no experimental evidence supporting such situations and it is difficult to imagine there ever being such evidence.

Although space must be dealt with relationally, it is important to remember that space is an essential element in social science analysis. It is intimately involved in the processes of identification and explanation. All substance laws have spatial information and explain spatial conditions. In fact, the only way that hypotheses can be tested or that laws can be confirmed and applied is by finding out if the facts accord with the hypotheses, if the facts are where they are supposed to be. To ignore spatial relations is to run the risk of being non-empirical.

A social science based on a physical science model and on the relational concept of space would integrate space, substance and time, in such a way that all three, though conceptually separable, would be linked in terms of definition or identification, and explanation. The practice of social science would involve one in addressing both the spatial and temporal aspects of substances; and in a mature social science, as in physical science, there would be no 'spatial' or 'temporal' or 'substance' disciplines. Some generalisations would include more spatial terms, and some more temporal, but as disciplines, there would be a balanced approach to space, time and substance. Theories of social science would include congruent and contiguous relationships as well as the interrelations between the two. Those questions which, in the extreme, present too great a separation of space from substance will not find expression in such a science. They either represent a misconception of what a science of human behaviour would be like, or they are about conceptions of space within other modes of thought.

Deviation from the model

Disciplinary emphases

These conclusions pertain to a higher state of development in the social sciences than now exists. The social sciences do not yet have explanations like those of the physical sciences and as a consequence, the end points of the space–substance axis are not conceptually recombined. A manifestation of the 'underdevelopment' of the social sciences with respect to the model is found in the manner by which areas of specialisation are defined. Areas of specialisation reflect what are thought to be significantly interrelated facts. In very rough terms, political science examines political facts, economics, economic facts, sociology, sociological facts, and so on. There is, however, considerable concern that the facts within these disciplines are not sufficiently interrelated to lead to significant scientific analyses and that the facts are not sufficiently independent of one another to be studied as separate subjects. To have a firmer idea about what constitutes a significant set of facts we would need to have some laws or theories connecting them. In the absence of these, the disciplines have little else to rely on for a definition of their field and boundaries than tradition.

Equally revealing is that areas of specialisation or emphasis have developed from the tripartite division of things into substance, temporal and spatial. There are disciplines which emphasise research in the substance, temporal or spatial attributes of things. No discipline, even in its most elementary stages, includes only one of the three, but most have a tradition of emphasis. History, for example, is about the past. Facts are examined as they form events and as they constitute an era, or epoch, or give character to a segment of time. Geography is interested in the characteristics of facts on the earth's surface as they are spatially and areally interrelated. The discipline is concerned with the face of the earth. The social sciences, in general, are interested in the systematic associations of substances in their respective fields. They do not explore the great breadth of space and time as do geography and history.[14]

Of course, within each discipline there are secondary emphases which tend to belie the primary concerns, and bring space, time and substance more into balance. For instance, the spatial element is explored in regional history, in regional economics, in ecological sociology. The principles of social science are involved in political, economic, urban and social geography, and time is included in historical geography. Yet, these secondary emphases do not erase the primary ones which make the cores of the disciplines distinct. Specialisation in space, time and substance may not be unwise, but it does not correspond to the division of knowledge in the physical sciences or in what would be a developed social science.

A most important manifestation of the divergence between the model of social science analysis and space, and the practice of social science, is the tendency of some social science generalisations to ignore the spatial properties of the facts they subsume. Often such ignorance is manifested in hypotheses with congruent spatial observational instructions. This is the case because if no explicit mention of spatial relations are made, perhaps because of a lack of interest or knowledge of such relations, the hypotheses will nevertheless read as though the facts were congruent. If congruent statements occur because of the omission of significant spatial relations, then such hypotheses will eventually be disconfirmed. Their facts will not be congruent, and significant variables will be missing.

Recent efforts to become more concerned about spatial relations have led to the formulation of more contiguous hypotheses. In geography, both kinds of generalisations have always been present because of this discipline's long-standing concern with spatial relations. However, despite a concerted and well-balanced effort by some areas of the social sciences to integrate space, time and substance, these disciplines have not succeeded in making social science generalisations of the rigour of those in the physical sciences. Social science generalisations are simply not as convincing as those of the physical sciences and neither are their statements about spatial relations. The problems are most clearly etched when we compare our theoretical formulations·of space in social science laws with the actual use of space in classical social science generalisations.

Ideally, space and substance will be conceptually connected through the relational concept and there will be two general forms of laws expressing spatial relations, congruent and contiguous laws. Actually, generalisations of the social sciences do tend to follow the two forms, but their linking of space and substance is not convincing. Congruent spatial relations seem artificial because they rely heavily on pre-assigned areal units, such as political jurisdictions, to mould the spatial properties of social facts. Contiguous relations are suspect because, to break this mould, they de-emphasise the substance referents of the spatial terms. They often are not directly about reality because they assume uniform, undifferentiated surfaces.

Congruent generalisations

In theory, congruent relationships are as spatial as the contiguous ones. Yet, hypotheses with congruent instructions have seemed suspect to those who emphasise spatial relations. Part of the suspicion is about the motivation for congruent statements. A number of such hypotheses are congruent because of carelessness, omission or unconcern about spatial relations. That is, they are congruent by default. One of the factors contributing to the lack of concern with spatial relations has been the erroneous idea that theory means only abstraction. This belief has been popular in the social sciences in part because it was felt to be the most immediate way of avoiding the enormous amount of detail involved in human behaviour. Abstraction meant lack of concern with the specifics, to the point where the formulations were not statements about empirical relations, but rather philosophical positions and wishful thinking. In addition, there was the notion that the most powerful social science theories would emerge by analysing what was often referred to as the 'system as a whole'. This led to an emphasis on activities at the largest social scale – the national. Smaller-scale activities within the nation were not only thought to be less significant in their impact, but also to lead to the examination of more parochial concerns which would not interest the public at large or the scholarly profession in general. The smaller scale was also equated with detail and hence, the less theoretical.[15]

Suspicions are further raised when congruency is the result of directly ignoring non-congruent relations. Some have contended that non-congruent relations and spatial relations in general are not important. The economist Marshall made a classic statement of this position when he subordinated the interests of space to those of time.[16] Economists and others who shared the view that the spatial dimension was not important, tended to ignore it by adopting such fictions as the supposition that all of the significant facts were located at a single point in space, or that transportation costs from place to place were zero. The avoidance of spatial detail has not added to the accuracy of economics. The assumption that all activities are at a single point is not equivalent to the assumption in physics of a mass point. In physics, mass point is not used to omit spatial detail, but rather to gain it in the sense that more accurate effects of the separation of bodies can be determined by employing this assumption.[17]

Perhaps the most fundamental suspicion of congruence arises from the way social facts are assigned spatial properties. One of the major underpinnings of congruent relations is the reliance on a limited number of areal units, such as census areas and political jurisdictions, to serve as spatial demarcations of social facts. These units often force social facts into spatial moulds and impart to them spatial properties which they would not otherwise possess. The activities and effects of cities, for example, do not all end abruptly at the same distance to form a clearly demarcated boundary. The effects of a city vary with each activity: sanitation, police and fire protection are circumscribed by the geopolitical boundaries of the city; economic impact and pollution are not. Yet, statistics about a city are given location in terms of the geopolitical units. Even the family is statistically anchored to the location of the family abode.

Geopolitical or statistical boundaries which are established for one purpose tend to serve many others for which they were not designed, and often the boundaries no longer coincide with the geography of the activities. A city's tax base is determined by its geopolitical boundaries, but the people a city serves may come from beyond its political boundaries.

Because statistical data are often given for the aggregate or group, the precise characteristics of the individuals comprising the group is not recoverable. Sometimes aggregation occurs to protect the privacy of individuals, as when voting records are recorded by wards and not by individuals, or when parts of the census material is released only in an aggregated form. Sometimes aggregation occurs simply for economy or convenience. Regardless of the reason, the statistics refer to spatial units, and these become the location and extension of social facts. Moreover, these spatial units are relatively few in number. Hence, there will tend to be a greater degree of areal accordance among facts than might otherwise be expected. The existence of such units tends also to make information about the facts within the boundaries more precise than information about the facts that 'pass through' such areas. The census has more information about the activities occurring within the tracts and areas than for those outside them. By making information about the activities within the boundaries more detailed than those outside and those of the past, we tend to disrupt or divide continuous processes into an exaggerated *here*, versus a vaguer *there*, or *elsewhere*.

Despite their reliance on spatial moulds, accordant relationships are empirical and accordant generalisations are scientifically sound. However, many who have emphasised the spatial properties of things look for more than accordance. This is especially true of those, like geographers, who have emphasised the spatial relationships among things and who look to the physical sciences for a model of explanation and find there that spatial variables are frequently explicitly included in laws and theories. But to go beyond congruence in social science is difficult. Even when one is studying clearly non-congruent activities such as migration or diffusion, it seems easier at times to assume congruence or to present the data in congruent form. There are models of diffusion of innovations which ignore the paths by which information is transmitted and assume instead that the information is either present or absent, not accepted or accepted, at a particular place for a particular population.

The logistics curve, for instance, is used to analyse the diffusion of innovations within a population. The curve describes the frequently occurring situation in which, at the beginning of the diffusion process, there is only one person who has accepted the innovation. After a time the number of people accepting the innovation increases and at an increasing rate, and then the rate of acceptors begins to decline until only a few are left who have not accepted the innovation; and these may take some time to adopt it, if they will at all. This general process can be described by the following logistics equation:[18]

$$P = \frac{U}{1 + e^{(a-bT)}}$$

where P is the proportion of adopters, T is the time at some point in the diffusion process, U is the upper limit of adoption, e is a base of natural logarithams, and a and b are values that change according to that particular kind of diffusion. The spatial observational instructions of this relationship are congruent. The relationship contends that at any particular time this population as a whole can be assigned a value representing the proportion of adopters. Where these adopters are within the population is not specified.

Similarly, a clearly non-congruent relationship between the economic growth of cities and their hinterlands is often studied as though it were congruent. The Economic Base concept argues that the cities exist and grow because of goods and services they produce and sell to their hinterlands. The model states that the total activity in the city (TA) will be equal to the goods and services produced within the city and sold outside the city – the basic activities (BA), plus the goods and services produced and sold within the city – the non-basic activities (NBA). In symbols, $TA = BA + NBA$. The actual proportion between a city's basic activities and its non-basic activities, or the proportion between the increase in basic and non-basic activities produces a basic service ratio from which estimates of the growth of the city can be made.[19]. The estimates assume that all the facts are located in the city. Even though the BA refers to interaction beyond the city, the interaction value is assigned geographically to the city. All information about the spatial properties of the interaction is omitted.

The avoidance of the contiguous relationships in these and other generalisations is due largely to our limited knowledge of scientifically significant facts in human behaviour. Non-congruence means we must know the causal paths and their spatial arrangement and these must be included in generalisations about the process. But as we have already pointed out, much of human interaction is fragmented to the point where we are unaware of what the communication links are. Even if we were to look at interactions which we know so well and over which we seem to have some control, generalizations about the effect of distance are qualified.

Contiguous generalisations

Studies of the positioning of people in normal, face-to-face conversations have pointed out that the communication of information and the positioning of the conversants depend on combinations of senses. We may not only want to hear what another person is saying, but also to look at his gestures, at his eyes, mouth and posture; we may smell his breath and his body odour; and we may use the sense of touch. These may all be employed in conversation and the nature of the conversation varies in terms of the degree to which one or another of these channels are used. Each sense has its own spatial range in conversation. We cannot touch, smell, hear or see people who are too far away, and the distances beyond which certain communications are impossible vary for each of these senses. Moreover, if the characteristics of the medium change so too do the ranges. Fog, darkness or noise will affect these ranges.

We are all aware from personal experience of these ranges and their effects on conversation. We know perhaps more about the effect of distance in face-to-face conversations than in any other kind of human interaction. This is no doubt due to the fact that the behaviour is not fragmented and is within the control of the speakers. We can see, feel, hear and otherwise sense the outcome of our actions and we know the medium through which the information is transmitted. But even here there are unpredictable elements linking distance to conversation. The precise spatial arrangements people will assume varies idiosyncratically and especially from culture to culture. Hall, for instance, claims that in normal conversations Americans do not prefer to be bathed in another's breath, whereas Arabs consider the smell of another's breath an important element in conversation.[20] Cultural preferences for combinations of senses and the use of personal distances as symbols make our knowledge of the effect of distance on a process as well-known and personally controllable as conversation, speculative and uncertain.

If we turn from the small-scale interactions about which we all have personal acquaintance to the larger and more complex interactions between large groups of people over longer distances, we multiply the problems of connecting space to substance. We will take an example analogous to proxemics but on a larger scale. Instead of considering people conversing, consider pairs of cities interacting. As in proxemics, it is assumed that the separation of cities affects the interaction between them. As distance increases the interaction decreases. Of course we cannot test this relationship by moving cities as we can pairs of people. Therefore, any generalisations we develop would be based on comparing distances between pairs of cities to their interactions. There are a group of hypotheses generalising these relationships and they are part of the 'laws of social physics'. These are modelled after generalisations of Newtonian physics. The social physics hypothesis bearing directly on the interaction between pairs of places is called the Gravity Model and is isomorphic with the Newtonian equation for the force of gravity which states that the force, F, of attraction between masses m_i and m_j are directly related to their product and inversely related to the square of the distance separating them, or,

$$F_{i,j} = K \frac{m_i m_j}{d_{i,j}^2}$$

In social physics, the relationship states that interaction I between two places i, j is a function of their populations p_i, p_j and the distance, d, between them. Symbolically,

$$I_{i,j} \approx f\left(\frac{\text{pop}_i \times \text{pop}_j}{d_{i,j}}\right)$$

Originally, it was expected that distance would have the same general effect on all interactions and would be non-relational; as Warntz put it: 'space and time are to be recognized not just as cost-incurring external frictions, but rather as dimensions of the economic system and hence to be treated isomorphically in the rigid pattern of mathematical physics.'[21] Stewart agreed:

In any physical situation alteration of the power (i.e., the exponent of one for the distance term) would be a serious matter, not one merely of the choice of an adjustable parameter . . . . The 'weight' assigned to the people is not similarly critical; it can and must be adjusted to fit the observations.[22]

Not surprisingly, this non-relational view has not been supported by the evidence on interaction. Rather, the effect of distance depends on the particular types of interaction and on the channels of communication and to reflect this variability, the exponent of d is considered to be a variable to be explained by other equations. But this awareness of the variable effect of distance may not necessarily mean that it is understood to be a consequence of the relational quality of distance.

Despite the more flexible distance exponent, the model has had problems in conforming to the facts. Not only does distance have a varied effect, but also the 'populations' can be measured in various ways. Incorporating these issues results in a more flexible equation, such as,

$$I_{i,j} = \left[\frac{(w_i P_i)(w_j P_j)}{(D_{ij}^b)} \right]$$

which $I_{i,j}$ is the volume of interactions between places i and j, w_i and w_j are empirically determined weights, P_i and P_j are the population sizes of place i and j, D_{ij} is the distance between places i and j, and b is an exponent measuring the friction of distance.[23]

Formulations associated with the gravity model have also tended to look at distance non-relationally by assuming away the substances between places. Reilly's law of retail gravitation, for example, assumes that 'all things being equal, two cities attract retail trade from any intermediate city or town approximately in direct proportion to the square of the distance from the two cities to the intermediate town'. The breaking point between the cities that are in competition for trade in their intermediate area can be determined as follows:[24]

$$\text{breaking point from city } A = \frac{\text{distance between } A \text{ and } B}{1 + \sqrt{\dfrac{\text{population of city } B}{\text{population of city } A}}}$$

Attempts have been made to reintroduce substance as in the

Intervening Opportunity Model. In the context of migration, the substances would be the number of opportunities that one can encounter along a move. The effect of the opportunities on behaviour is that the number of people moving a given distance is directly proportional to the number of opportunities at that distance and inversely proportional to the 'number of intervening opportunities'. Knowing which places are going to be intervening opportunities means that we must know the path or route the interaction will take, and this is not necessarily a straight line, nor is it suggested in the model how it is to be determined.[25]

Hypotheses about non-congruent relationships of greater complexity seem to consider space even more non-relationally. Central Place Theory considers the spatial interrelationships of a system of cities as market centres, and fundamental to the theory is its simplified space. The theory assumes among other things, an unbounded, flat, uniform, surface in which movement is equally possible in all directions and in which transportation costs are a function of distance. The initial population of the place is distributed uniformly. Each consumer has the same knowledge, preference incomes and propensities to consume and the same demand schedules. They are living in a perfectly competitive market system and they are rational consumers, that is, they prefer spending less than more. There are a large number of buyers and sellers and the prices for the same goods are the same everywhere. In addition, all consumers are to be served, and by the maximum number of establishments. Given these assumptions, the cities will locate to form a hierarchy of hexagons (see Figure 3.1) and, according to one formulation of the theory, 'for every trade area of a given order, there are three trade areas of the next lower order'; and 'the distance between higher order centers is equal to the square root of three times the distance between the next lower order centers'.[26]

Similar assumptions about flat uniform surfaces are made in land-use theories. A classic example is Von Thünen's theory of agricultural land use around an urban market centre. Like Central Place Theory, Von Thünen's theory assumes rational people with perfect information. In the model they are located around an isolated market centre. The centre receives the produce of the farms from its hinterland and from nowhere else, and the farmers ship to no other centres. The surface on which the farmers are located is completely uniform with regard to soil, moisture and any other factors affecting

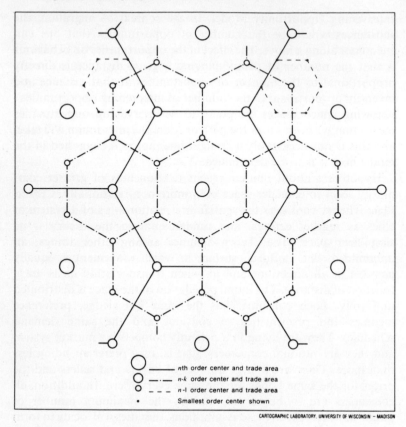

nth order center and trade area
n-k order center and trade area
n-l order center and trade area
Smallest order center shown

FIGURE 3.1   *Central place hierarchy*

agricultural productivity. There is only one mode of transportation to the market and all points equidistant from the market are equally accessible. Transportation costs vary with distance to the market and vary also from crop to crop. Prices at the market are the same for all farmers and vary for each crop. The transportation costs are borne by the farmer. Given these factors, concentric zones of agricultural activity can be expected around the market, wherein the more intensive agricultural activities and the more perishable crops are located closer to the market and the less intensive agricultural activities and less perishable crops are located farther from the market. The precise number of zones and their extent depend on the number of crops, their prices at the market, the transportation costs and the rent.[27]

To test the conclusions of Central Place Theory and Von Thünen's theory, we need conditions approximate to those assumed in the theory. We need a relatively closed system in which the inhabitants are rational decision-makers with perfect information living in a flat isotropic surface. Several areas on the earth's surface tend to approximate to the consequent conditions of the theories although no areas are exactly like this. For instance, approximate hexagonal arrangements are found in Southern Germany (see Figure 3.2) and in Szechwan Province, China, and concentric, though not circular, zones of agricultural activities are found at all levels, from city to nation. But these are of the consequent conditions. The problem is in determining whether these were caused by the antecedent conditions discussed in the theories. Testing such relationships brings us back to the problem of significant substances which were assumed away in the hypotheses.

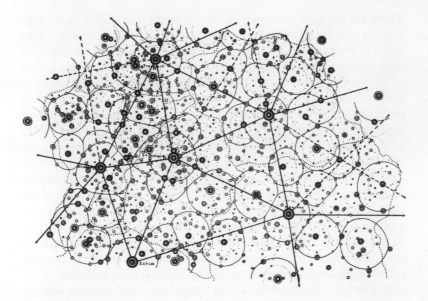

FIGURE 3.2 *Distribution of central places and hinterlands in Southern Germany (after Christaller)*

The examples above are the ones closest to the predicted consequent conditions. Most agricultural and urban systems do not look like this at all. How imperfect the actual shapes are on the ground is difficult to judge because there is no clear measure of shape.[28] Even if there were, we would have a difficult time determining what non-isotropic substances contributed to the spatial distortions because of the difficulty in measuring shape. Most tests of the consequent conditions of Central Place Theory concentrate on the average spacings among the places and on the average sizes of the hinterlands. While some systems of cities correspond to what would be expected from Central Place Theory, there has been considerable difficulty in connecting these consequent conditions to such antecedent conditions as perfect rationality, minimisation of distance, clear hierarchies of functions and so on. It should be noted that despite these difficulties, Central Place Theory and other location theories are as elaborate and accurate as social science generalisations come, and have been extremely popular in disciplines outside geography and have been used in planning.[29]

These are but a few examples of the problems involved in recombining contiguous formulations with substances and these difficulties bring us back to the problem of the spatial observational instructions in social science and the general problem of combining the end points of the space–substance axis. In general, congruent hypotheses in social science may tend to be suspect, because they may appear not to pay attention to spatial relations. They may be congruent because of aggregation or default. On the other hand, contiguous laws have tended to incorporate distance and other variables at the expense of connecting them with significant substance referents. Problems with both types of laws have led some to mistakenly believe that the effects of space on behaviour could be studied best only through yet another class of laws, namely 'spatial laws'. These would be the kind which would answer spatial questions. What exactly the spatial laws were to be like was not always made clear. A mild view suggested contiguous laws which do not adhere to a relational concept of space. These were often called static, morphological laws with spatial variables.[30]

A more extreme case held that spatial laws are like the laws of geometry and could offer 'pure' spatial explanation and prediction. These laws would be able to explain and predict the forms that things possessed from general principles about spatial relations. Like geometry, such principles would be unaffected by the particular substances which manifested them.[31]

Such excesses were attractive to some because they held out the promise of a kind of explanation that would be able to embrace social and physical facts equally by embracing their spatial forms and properties. These general spatial laws, it was hoped, would serve as a unification of geography's concern with spatial relations. Not only would physical and human geographers share the same questions, but they would also be able to rely on the same explanations. Such diverse phenomena as stream morphology and highway networks were to be united under the same generalisations. Such excesses were prompted also by an over-reaction to the lack of concern in the other social sciences with spatial relations. But most importantly, they stemmed from a poor understanding of the way in which explanation in the social sciences would work if in fact there were laws of human behaviour. It is the same kind of misunderstanding of the role of space in social science explanations that gave rise to the underemphasis of space in other social sciences.

Ignoring spatial relations or conceiving of them non-relationally will hinder the discovery and confirmation of social science generalisations. However, an awareness of and adherence to the role of space in social science explanations by no means assures the success of social science. Problems of understanding human behaviour scientifically are legion and the relationship between space and substance is only one of them. But space is a sensitive indicator of the success of explanations. Because social science explanations are weak, so too is our knowledge of the significant relationships between space and substance.

4
Social Science and Subjective Meanings of Space

Laws of social science would conceptually recombine the end points of the space–substance axis through the relational concept of space. In the absence of such laws, two perspectives closely related to the classical social science model – behavioural geography and chorology – become extremely important approaches to the link between space and substance. The behavioural approach has occurred in response to the weaknesses of the classical social science perspectives. Chorology, however, existed long before there was such a classical view. In fact, it has been characterised as the traditional form of geographical analysis.[1] Nevertheless, the logic of chorology and its recent role in geographic thought can be best represented as a modification of the classical social science perspective.

Chorology and behavioural geography approach the relationship between space and substance by moving farther into the subjective direction along the objective–subjective axis. They attempt to link space and substance by creating views of space that are more subjective than are found in the classical social science view. Chorology and behavioural geography arrive at this link and the subjective view of space differently. Chorology is an approach to regional analysis and synthesis, and results in the concept of the *specific region, place* or *area*. We will use these terms interchangeably although they sometimes connote different geographical scales. Chorology assumes a Euclidean description of space and links substances to it by assuming the spatial properties of acts stem from their location in this space. It differs from the classical model in that it links space to substance by relaxing the scientific criteria of significance. This relaxation makes place more specific and subjective. The sense of place that comes from chorological analysis is similar to the sense of time (such as era, epoch or biography) that comes from the study of history.

The fundamental orientation of the behavioural approach in geography has been to consider alternate descriptions of the spatial properties of things in the form of cognitive distances and cognitive and mental maps. These descriptions of spatial relations are attempts at capturing the subjective meanings or definitions of the spatial properties of things or even of space itself and translating them into objective terms. These descriptions are then expected to be significantly related to substance.

Chorology indirectly, and behavioural analysis directly, attempt scientifically to analyse the subjective meanings of space and place; and analysing the subjective is the fundamental task of social science. Neither of these orientations, however, penetrates very deeply into the subjective. The path to feelings about space, the extent of the path that the sciences need to traverse, and the obstacles confronting the social sciences along it, will be seen more clearly when we consider the problems of scientifically analysing the subjective import of art. Art is the mode in the sophisticated pattern which most clearly involves the subjective end of the axis. In art, space and substance are conceptually separated and given significance differently than they are in science. As we noted in Chapter 1, the visual arts use substances such as lines and colours to create the illusion of space. In the non-visual arts, such as poetry and prose, words are used to create the illusion of space. In both the visual and non-visual arts, the illusion of space is the conceptual separation of space. The meaning or significance of the illusion depends on the feelings it symbolises.

Chorology

Facts are located in space and their areal delimitation defines geographic places, areas or regions. The areal delimitation of a well-defined fact or set of facts such as those which are instances of scientific concepts, is one meaning of *generic* place, area or region. An area of 30–40 inches of rainfall per year is a generic area. With regard to precipitation, it is like any other area of 30–40 inches of rainfall. Of course, there are other facts located within the area which may not be alike from region to region. One 30–40 inch region may be a corn-producing area, another may be a wheat-producing region; one may be densely inhabited, another sparsely inhabited. But regarding the location of rainfall only, all 30–40 inch areas are, by definition, alike. They are regions of a general kind.

Another meaning of *generic* place, area or region refers to the areal delimitation of a system. (A system is a set of facts all of which are significantly interrelated.) Laws and theories identify systems; they identify significant facts and their interrelationships. The relational concept points to the kinds of spatial configurations of systems that one can expect. Facts in a system will be either spatially congruent, contiguous, or some combination of the two. Areally delimiting a system identifies a generic place, with the reminder that systems on the surface of the earth are open, and hence, the spatial delimitation will always be only partially inclusive of the system. Areally circumscribing the general relationships between rainfall, social-cultural artifacts, and crops, identifies generic farming regions or farming systems. Such generic regions would be the consequence of observing and testing scientific generalisations about farming. The generic place would be the spatial accompaniment of accurate description and scientific explanation.[2]

There is, however, another meaning of place which comes about when we include and even emphasise the difference among places as well as their similarities. We know that generic regions mean that certain things are in association at a place, as in corn-producing regions, wherein we will expect to find certain kinds and amounts of accompanying factors, such as soil type, temperature, precipitation, length of growing season, cultural attitudes, values, transportation infrastructures, and so on. In addition to these expected factors in each corn region, other unexpected factors will be found, and these will vary from region to region. Some corn regions, for example, specialise in cattle, others in hogs; some are near and others far from the market. These unexpected and different factors and associations affect the character of the places and make each generic place different, up to a point–sometimes slightly and sometimes very different. Of course they cannot be completely different because they still share the common attributes and relationships that make them generic. The incorporation of these 'exceptional' factors and their interrelationships within a place, area or region is analysed through *chorological syntheses* and the result is a *specific* area, region or place. Incorporating the expected and the unexpected is what occurs when we explain the character of something as complex as a neighbourhood, a city or an agricultural region.[3]

The specific place, like the generic, is the areal circumscription of interrelated facts. But the facts and interrelations circumscribed in chorological syntheses, because they include unexpected associations, are, by definition, not entirely subsumed within social science generalisations. Chorological analysis goes beyond scientific generalisations in connecting space and substance, but its procedures are best expressed in relationship to such generalisations.

We can think of a specific region or the specificity of place as coming about when the generic fails. Take the simple example of water, temperature, pressure and ice. We expect that ice will form when water falls below 32° Fahrenheit. The areal circumscription of the facts in this example constitutes a generic place or region. If we come across some water that does not freeze at 32° Fahrenheit, we look in the vicinity to determine why. We begin to think of this exception in terms of the place.

The factors which may contribute to the water not changing state would be significant and yet would be beyond the original generalisation. But herein lies the problem. For we cannot be sure that the factors contribute to the water not freezing unless we are in possession of other generalisations within which these facts are subsumed and connected to the freezing point of water. For instance, we know that salt or other chemicals in the water would lower the freezing point because we have generalisations and theoretical knowledge which support this. If the water in fact contains salt, the place then is not what we originally expected it to be (it is specific with regard to the first generalisation), but our surprise can be explained away by another and more comprehensive generalisation which makes the place once again generic. However, we may not be sure about the reasons for the change in freezing point. Either we may not have this second generalisation or else it and other generalisations do not explain the conditions in this place. Yet we may think that some of the facts in the vicinity may contribute to the change in freezing point. In such cases our synthesis is less certain and the place becomes more specific. Such uncertainties are far more prevalent in the social than in the physical sciences. In fact, they are endemic in the former. The facts which need to be explained in social science are far more complex than those handled by the few generalisations in which we do have confidence. It is not surprising then that chorological syntheses are part of social geography and not physical geography and methodological discussions on the subject use social science examples.

Hartshorne, who has written the most about the specific place and believes it to be the fundamental concept of place in geography, discusses the specific place in terms of the failure of or the limitations of the generic or nomothetic (that is, generalisations). Take, for example, the generic relationship between annual variations in corn production and annual variations in rainfall. He points out that simply examining the relationship in terms of areal association of these facts is not going to lead to the specific region or place unless there are marked variations in this relationship. Thus, 'the fact that variations in rainfall in Nebraska have a greater effect on corn yields, than the same degree of variation in Pennsylvania, is of geographic concern', because it leads to the analysis of the specific place.[4] 'Geography [as the study of specific places] is concerned with [such factors as] the marked areal variation in corn production, since this presents a part of the total areal differentiation in which it is associated on the one hand, in its relations to differences in climate, soil, relative location, or cultural conditions and on the other hand, in its relation to differences in the total crop livestock element-complex, the character of barns, the presence of grain elevators, etc. . . .'

We use systematic knowledge to compare and contrast, to integrate and differentiate, but insofar as our systematic knowledge is accurate and true we have not yet reached the characteristics of the place that make it specific or unique. In the maize example, if there were a correspondence between corn yield and rainfall, it would qualify as an empirical generalisation. It would explain and predict the locational aspects of corn yield as being accordant with the location of a specific amount of rainfall. This would be systematic, lawful knowledge, and the delimitations of the accordant relationship between corn and rain could be thought of as generic, rather than specific, places.

Specific regions come about when there are exceptions to the generalisations (or when the generalisations are inadequate) and when it is supposed that at least some of the significant factors which were omitted in the law or hypothesis are located in the vicinity of the facts to be explained. This supposition about looking in the vicinity follows from the fact that the interacting facts will have to be linked through space by intervening substances. Hence we should be able to observe, in the spatial vicinity of one variable, some of the other significant variables which may have been omitted from the hypothesis. In this respect, something like what Lukermann termed 'local conditionality' can be a realistic procedure for scientific observation and discovery, although one must remember that the final variables may be far as well as near.[5]

But looking in the vicinity for the variables or links to other variables far away requires that we have some idea of how they are significantly interconnected. In a traditional scientific sense, this means either that we have a new hypothesis or set of hypotheses to incorporate these 'other' concepts, or else that we modify or correct an old one to include them. In either case, the syntheses would result in a generic place, albeit a more complex one. To argue that the synthesis depends on something else is to make it less scientific or less scientifically significant (albeit the nomothetic criteria of significance are not adequate by themselves). And what else it could depend on opens up enormous ranges of opportunities for different non-scientific methodologies. But chorology has traditionally eschewed clearly non-scientific methodologies. Chorological synthesis may be done aesthetically, but art and holism have not had explicit parts in chorological methodology. Although chorology goes beyond the study of generalisations, it is scientific in its accurate descriptions and orderly presentations and in its adherence to facts. To model this kind of analysis we need to find modes of syntheses which are looser than scientific laws or hypotheses and yet which do not completely abandon the scientific criteria of significance.

Disciplines which hold that the discovery of laws is their first and foremost obligation tend to study facts which are subsumable as instances of scientifically significant concepts. Given the paucity of significant concepts in the social sciences, such a commitment might exclude a great deal about which we are curious. Our inability to subsume this vast residue of facts under the nomothetic obviously has not deterred everyone from attempting to analyse them. From every social science we have innumerable examples of studies which do not test hypotheses and yet which attempt to explain them, and in a fashion which is committed to facts and to objectivity. Compared with the discovery of laws and interests in hypotheses testing, such attempts are often 'long on facts and short on theories'. To the extent that they are offered as explanations they do not have the generality or predictive power that are the hallmarks of the classical social science models and their analytic structures are rarely presented within a specific methodological context.

Describing what precisely constitutes the systems of analysis in such discourse has been of considerable interest and debate among philosophers and disciplinary methodologists. The model which may best capture such synthesis is something along the lines of the *explanation sketch*.[6] Explanation sketches seem the most versatile and the closest to the nomothetic of the more familiar models of analysis. They have been offered as models to describe what might be meant by facts explaining other facts in normal conversations about the causes of human actions and in such discourses as history, which are long on facts and short on theory. Explanation sketches are incomplete generalisations or hypotheses. They refer to the reconstruction of possible general relations that are invoked in empirical studies which claim to 'explain' facts by other facts, but which have offered few if any explicit generalisations to subsume these facts. A sketch 'consists of a more or less vague indication of the laws and initial conditions considered as relevant, and it needs "filling out" in order to turn into a full-fledged explanation. This filling-out requires further empirical research, for. which the sketch suggests the direction.'[7]

If, for instance, Johnny failed an examination and we say in normal conversation that Johnny failed because he did not study, and if we show that Johnny in fact did not study, we may accept this explanation of Johnny's failure as plausible, because we may be assuming there to exist a plausible hypothesis which has not yet been fully articulated or tested but which, if it were, would likely state that, given certain circumstances, studying contributes to the passing of examinations. Accepting such an explanation of Johnny's failure suggests that we are appealing to such an implicit hypothesis or explanation sketch. If we do not accept this particular explanation but offer another, for example, that Johnny is not intelligent enough to pass the examination, we are then appealing to yet another sketch, that is, that under certain circumstances passing an examination like this requires a certain IQ. In general, the plausibility of the discourse depends on the veracity of the facts, and on the 'validity' of the invoked laws, hypotheses and explanation sketches which together would make the facts significant.[8]

Historical explanations have been offered as examples of explanation sketches. Historical analysis most often does not test hypotheses or even explicitly employ them. Yet such discourses are factual and the facts are used to explain other facts. Moreover, historical analysis is judged on the merits of its 'explanations'. It is argued that the facts are significantly related in such discussions by strings of explanation sketches. Asserting, for instance, that the protection of property was a determining factor in causing the American Revolution, and then 'proving' that there were colonists who felt that they could do financially far better alone than under the yoke of Britain, is a plausible explanation (assuming the facts are correct) because the readers of history, as well as the historians proposing it, assume without explicitly stating the hypothesis and all of its preconditions and without testing it, that, in general, economic concerns can lead to revolution. Because the facts in such explanations are only tentatively connected, historical analysis not only produces an explanation whose criteria of validity are more relaxed and whose explanations are therefore more subjective than those of classical social science, but it produces a different sense of time than is portrayed in the units of minutes, hours and years which so often accompany classical social science. Rather, historical analysis creates distinct and incommensurable temporal units such as lifetimes, eras and epochs.

Chorology is similar to history in its forms of synthesis; and like the conception of time in history, the conception of space in chorology is different from the commensurable units of lengths, shapes and areas. Chorological syntheses can be said to use explanation sketches in addition to whatever scientific generalisations may be available. Explanation sketches, as incomplete nomothetic statements, include the spatial observational instructions we identified within scientific hypotheses and hence connect the end points of the space–substance axis. The facts discussed in such sketches would be interconnected in space either congruently, contiguously, or in some combination of the two. Because of the geographic interest in spatial relations, chorology has given these linkages especial attention. The concentration on spatial linkages is important, for in the absence of social science laws, the establishment of spatial links among the facts demonstrates that at least one necessary condition of scientific significance has been satisfied. This, in and of itself, may make the facts seem plausibly connected in a causal sense. But the reliance on explanation sketches makes the identified spatial linkages still only tentative routes in possible causal chains, and the lack of comprehensive theory leaves gaps in these chains. The use of explanation sketches in chorology allows the syntheses of facts too complex to subsume under clear social science generalizations and shows how their interrelationships form a place. Since such syntheses, with their use of explanation sketches, can be no more than plausible, tentative and partial, the facts of such a discourse collectively are relatively unique in their association in space. In their areal delimitation, they constitute a specific region, area or place.

Chorology emphasises spatial linkages. These linkages, as well as the place they define, are specific because they are not explained or predicted by laws. Rather, they are subsumed in explanation sketches. Part of the tentativeness and subjectivity which accompany explanation sketches is transferred to the spatial relationships. By emphasising a necessary criterion of interconnection (that is, spatial linkages), while not specifying in as great detail other equally necessary criteria, spatial relations and place in chorology appear to possess causal efficacy in and of themselves. This appearance results from the combination of emphasis on spatial relations and explanatory incompleteness. The shift in meaning of space and spatial relations that comes from chorology is comparable to the change in meaning which time undergoes when we shift from the context of hypotheses and their temporal variables, to history and its characterisation of the temporal connections among facts as distinct periods of time which seem to possess a character and causal efficacy uniquely their own. We say someone is a product of his times, as we say someone is a product of his place. Strictly speaking, these are only locutions about facts in space and time. But these locutions indicate a different, and unavoidable, evaluation of space and time which comes from incomplete scientific syntheses.

Modelling chorology in terms of explanation sketches is another way of saying that chorology, albeit reluctantly, is unavoidably to some degree subjective. It is, though, a controlled subjectivity which involves critical scientific judgement. The chorological tradition has adamantly maintained that chorology is to be factual, orderly description and evaluation, and that chorological synthesis should not involve aesthetic modes and metaphysical holism. Nonetheless, its relaxation of the nomothetic strictures introduces subjective elements in the syntheses, which in chorology extend to the understanding of space and spatial relations and to their connection to substance. By progressively relaxing the strictures of the nomothetic we can imagine backing into ever more subjective analyses of space and place until we are completely beyond the realm of science and rely on personal evaluations of spatial relations or subjective evaluations such as aesthetic ones. Chorology can thus be seen as a stepping stone in the direction of increasingly subjective evaluations of space. But its practitioners have kept it linked closely to the sciences. The best examples of chorology, like the best examples of history, are not flights of fancy or creations of art, although they may have aesthetic appeal. Rather, they are descriptions and analyses of complex facts and relationships in the most scientifically plausible way.

While chorology modifies the meaning of space and place by introducing subjective elements into the synthesis, it does not lead directly to an examination of subjective senses and meanings of space and place. The subjective meanings of space have been more directly confronted in geography and in the other social sciences in what we will call the behavioural approach.

Behaviouralism

Dissatisfaction with the attempts to find laws of human behaviour has led to a change in emphasis in the social sciences, a change which directed research closer to the subjective aspects, by approaching behaviour from the viewpoint of the decision-maker. The thrust of the change was expressed in the idea that people behave according to the way they perceive the world, not simply on the basis of the way the world is. Knowing how the decision-makers see the world (knowing their psycho-milieu, as it is called) will help to explain and predict their behaviour. Moreover, seeing the world through the decision-makers' eyes, it was felt, will present a deeper understanding of human actions. This orientation is known in geography and other social sciences as the behavioural approach.[9]

In terms of the space–substance axis, we will focus on that part of behavioural geography which attempts to describe the location and extension of things in space from the viewpoint of the decision-makers with the expectation that this new description can be *significantly* related to substance. This research approach criticised what seemed to be the erroneous assumptions of social scientists that the significant description of spatial properties in particular and of variables in general could be obtained from evidence and data as it was presented to them by the physical sciences, as it appeared to them as 'objective' observers, and as it was collected by conventional data-gathering agencies which asked limited questions and which often aggregated the results. Furthermore, the increasing finances available to the social scientists in the 1960s and 1970s and their access to data-processing facilities made it possible for researchers to ask their own questions, to collect their own data, and sometimes to analyse the results at a disaggregated level. Attempts to understand the decision-makers' point of view by eliciting such subjective factors as attitudes, values and preferences (through such devices as multi-dimensional scaling, semantic differentials, grid reportory tests etc.) placed the behavioural approach close to the psychological sciences. This was seen as an advantage because of the notion that individual psychological analysis is more fundamental than the analysis of groups and their characteristics; an idea that has been buttressed by the concept of reduction.[10] But the overriding intention of behaviouralism has been to make translations/correspondences of subjective states of the decision-makers into/with objective states.

As in the classical social science view from which it evolved, there exists in the behavioural approach both under and overemphasis on spatial characteristics. These in turn have created problems of scientific scope and validity similar to the one discussed in the traditional approach. In political science, for example, analyses of legislative processes began to include studies of the motivations and 'perceptions' of legislators as well as the more traditional analysis of institutional rules, regulations and procedures.[11] While such studies expanded understanding of the decision-making process, they did not concern the spatial properties of the facts at a scale and in that detail which most geographers and others with spatial bents are interested in, nor did they question the significance of the spatial properties of the facts they did explore. Thus, generalisations from these studies relied on congruent relations and usually by default.

In part to compensate for such underemphasis, and to expand the analysis which traditionally included explicit examinations of spatial relations, the behavioural approach in geography emphasised the cognition of space and its properties at various scales. The approach assumed that space and behaviour could be conceptually linked by viewing space and its properties from the eyes of the decision-makers. Perceived or, as it is referred to in the literature, cognised spatial relations may contribute more significant measures than physical spatial ones.

The most frequently measured cognised spatial variable is distance. From our everyday experiences we know that our cognitions of physical distances often deviate from real distance. It seems to shrink and expand according to our feelings about the origins, routes and destinations. We may feel that the distance we travel to a place we do not like is longer than the return trip to a place we enjoy. Pleasant scenery or company may 'shorten' a journey, and monotony and bad company 'lengthen' it. Attempts have been made recently to quantify and 'objectify' such subjective observations and to determine the quantitative relationships between perceived or cognised and real distances.

A general relationship between the two distances was sought as a first approximation. Such a relationship is possible but unlikely from a relational viewpoint, because the cognised distances would depend on the actions and substances, of which there are innumerable kinds. It is not surprising then that the search for such a generalisation has produced conflicting results. Some have found the relationship to be described by a linear function[12] and others by a power function.[13]

On a more piecemeal level different measures of cognised distance may be due in part simply to the different techniques used to elicit the information. Simple mileage estimates, direct magnitude estimates and ratio estimates have all been shown to produce different estimates of cognised distances.[14] These differences are compounded by the fact that the results vary according to geographic scale.

Most importantly, from the relational view, the association between perceived and actual distance depends on the individuals, their goals and routes. Unfortunately, little is known about the effects of such factors. Yet their impact is felt over again in the problems of interpreting cognised distances.

In addition to experiments on distance functions there have been studies on the cognition of directions and shapes. Studies which include these, as well as distance, are usually referred to as *cognitive* or *mental maps*. They, like the distance experiments, are based on a variety of methodologies, refer to a variety of geographical scales, and tend to omit contexts and goals. Often subjects are asked to draw maps from memory of a geographical arrangement such as a room and its furnishings, a neighbourhood, a city, a nation or the world, and the results are compared with the original distribution to determine the distortions and perhaps their causes.[15]

In cases where the differences are not easily seen, evaluation of cognitive maps would require refined measures of shapes and distortions. A most general (and as yet unattained) description of the distinction between the real and cognised map would be to find the mathematical function which will transform the real patterns into the cognised ones.[16] The nature of this transformation would depend on the goals and substances which are distributed in space, and an analysis of the distortions would include relating the distortions to the goals and substances which the maps portray. Given the complexity of the substances, it is unlikely that there will be a single set of transformations of real patterns into cognised ones, just as it is unlikely that there is a general relationship between real and cognised distances.

Often the causes of distortions in cognitive maps and their significance for human action are not the primary objective. Although some attention may be given to these issues, the works tend to present the maps as ends in themselves. These are offered as intrinsically significant in their graphic presentation of subjective viewpoints. This is especially true of fanciful maps which depict a view of some imaginary person like *the New Yorker's view of the United States*. They are expected to be arresting and suggestive without exegesis.

Although such studies are intriguing, the question of their scientific significance is open if the relationships between cognitive spatial relations and behaviour is not more precisely established. One way of judging the scientific significance of these measures would be to substitute cognitive spatial measures for physical spatial ones in social science hypotheses which explicitly include spatial terms, like distance, and see if the explanation or prediction is improved. If so, this particular cognitive distance measure would be a more significant one than physical distance. For example, to explain the selection of supermarkets in a neighbourhood, Cadwallader used a model which included the price of goods, parking facilities and the physical distances to these stores. He then used a similar model in which he substituted a measure of cognised distance for real distance. Comparison of the results showed that the model using a cognised distance measure contributed to more of an explanation than the one using a real distance measure.[17]

Other experiments like this would present a clearer picture of the significance of cognitive spatial measures. But they also would require clear and quantifiable hypotheses about human behaviour, of which there are few. If we exclude the hypothesis testing approach for the moment, the primary means of justifying cognitive maps seems to rest on the most general belief that allowing us to 'see' the world or part of it from the eyes of the decision-maker is self-evidently edifying. We say 'seems' to be because the issue has never been fully addressed. The belief that cognitive maps are edifying would have support if it could be shown that decision-makers employ such maps when making decisions; that is, they visualise the environment or possess information which can be reconstructed by the researcher in the visual form of a map. While leaving a great deal about decision-making unexplained, this assumption about the decision-maker's use of cognitive maps obviously would integrate the cognitive maps with decision-making processes.

However, researchers have equivocated about whether the cognitive maps are in fact visualised by the decision-makers and whether decision-makers use relationships about the physical world which can be represented in map form. As though to allow for all possibilities, some definitions of cognitive maps are far more general than that of a literal map. Cognitive maps have been defined very broadly as 'schemata' or a 'cognitive representation' of the environment, which need not be in map form.[18] These meanings are extremely broad; for according to them, any model of the mind can become a cognitive map. Despite the extremely open-ended definition of cognitive map, the research designs and their products are much more specific. They point to visual images of spatial relations or to information about the environment that can be portrayed in visual form. Therefore, it would seem that the overriding meaning of cognitive map that most researchers have has to do with either the internal visualisation of reality by the decision-maker, or to possession by the decision-maker of information about the environment that is observable and presentable in map-like form. As we have said, linking visualisation of the environment or the use of knowledge that seems to be stored in map-like form with decision-making would indeed lend significance to cognitive map research.

What then of the suppositions that cognitive maps are used in decision-making, in the sense that the decision-maker needs to visualise the spatial properties of the environment, or to possess information about spatial relations which can be reconstructed by the researcher in visual map-like form? One of the difficulties in determining the significance of cognitive maps for behaviour is that many cognitive map experiments were done in a non-relational perspective. They were designed to portray images of places, shapes and distances while leaving the substance content to a minimum or discarding it completely. Hence, there is often a mismatch between the spatial concepts, that is, the relationships in the maps such as the places, the distances, the directions, and so on, and the substances on the ground; so that we do not know which substances caused which distortions and which distortions affect behaviour. Even if such mismatches are not serious having people make such maps does not mean they use them or the information they contain. A careful examination of the data from experiments on perception of distance indicates that respondents may not have visualised or even possessed the information to visualise the map-like relationships among the places about which they were asked. To be able to visualise spatial relationships assumes that the distances between places be com-mutative and transitive. Commutative means that the distance from point *a* to point *b* is equal to the distance from point *b* to point *a*. Transitive means that if the distance between *a* and *b* is greater than the distance between *b* and *c*, and *b* and *c* is greater than *c* and *d*, then the distance between *a* and *b* is greater than the distance between *c* and *d*. To repeat, transitivity and commutativity would be expected in a geometric relationship that can be visualised.

Cognitive distances based on paired comparison and direct estimation techniques reveal quite a number of respondents who give answers which are non-transitive and non-commutative.[19] These answers may be due to estimation errors, or more likely, they may show that respondents, when asked about distances among places, were not visualising the places in map form or did not possess information which can be used to construct a map.

Fragmentation of our actions and the use of technological devices which remove the paths and networks from our view and purview make visualisation of many of our actions difficult if not impossible and unnecessary. Decision to converse with someone on the telephone need not call to mind the location of anything except perhaps the location of the phone. Even this may not be necessary if the person is familiar with his surroundings. Familiarity allows us to accomplish many complex acts of locomotion without imagining the spatial arrangements of things. In fact, imagining such arrangements may hinder our movements. The experienced pianist does not picture the location of the keys when playing the piano, nor does the experienced typist recall the keys when typing. If the pianist or typist were to visualise the keys they would not be able to play or type effectively.[20]

Imagining actual spatial relations is a mental reconstruction of experience which requires considerable effort and concentration, so much so that we must stop to think about them. As Yi-Fu Tuan points out, such images or maps may be involved in giving directions to some one who is lost. Telling someone how to go somewhere may require imagining the route, especially if we were never asked this question before. Cognitive maps may also be used as mnemonic devices. We may be able to recall to whom we spoke at a party by remembering where that person stood in the room. Another area in which mental maps occur is the imagining of spatial relations in the form of rehearsing bodily movements. Such mental rehearsals can increase the efficiency or effectiveness of our movements.[21] Mentally rehearsing the movements of dart-throwing, basketball free-throws, pole vaulting and other activities may increase our performances in these activities.

From the geographic perspective, the most important realms in which cognitive maps are known to occur is in cosmographies and in the images of new places, *terrae incognitae*, and in Utopian thought. Such images of place are significant in the self-evident and generally accepted sense that they represent a picture of what people think. Cosmographies are excellent sources of cognitive maps. Descriptions of the world which most radically differ from factual Western accounts are found in the cosmographies of primitive societies. Such images not only differ from actual geographic arrangements, but, as we shall discuss later, they are evaluated differently. They are often seen through a magical–mythical mode.

We need not, of course, go to non-Western societies to find enormous distortions in or magical –mythical evaluations of the images of the world. A prominent medieval view, as cartographically depicted in the T–O maps, shows a markedly distorted conception of the earth (see Figure 4.1). These maps are pictures of the world drawn largely from Biblical sources. The east, rather than the north, is usually at the top of the map (hence the term orientation). The rivers Don and Nile form the top of the T and the stem is the Mediterranean. Above the Don and Nile is Asia. Europe is below the Don, and Africa, below the Nile. Jerusalem is at the centre of the world. As the Lord says, 'this is Jerusalem; I have set her in the centre of the nations, with countries round about her.'[22] Surrounding the world is the circumfluent and along the edges of the world we find the winds. Some of the T–O maps located such mythical features as Gog,

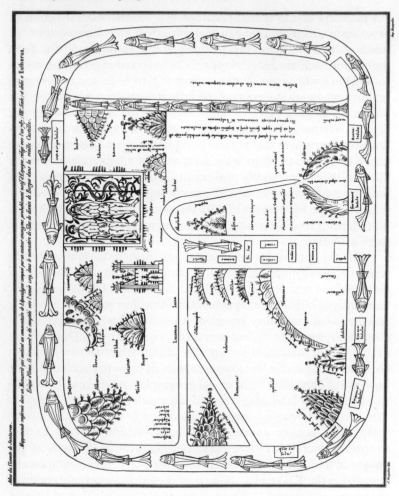

FIGURE 4.1   *Tenth-century rectangular T–O map*

Magog, the ark and the Garden of Eden.[23] Sometimes these maps
were not circular but rectangular, after the Biblical pronouncement,
'He will raise an ensign for the nations, and will assemble the outcasts
of Israel and gather the dispersed of Judah from the four corners of
the earth.'[24]

Closer to home were the fifteenth-century views of a water passage across the Atlantic to China, which gave rise to the 'discovery' of the new world; the legend of the seven cities of gold, which gave impetus to the explorations of DeSoto and Coronado; belief in the passage to India which influenced North American explorations; and general misconceptions about places which influenced decisions about staying or moving and affected how people adjusted to places.[25]

Images of spatial relations can be imposed upon the land at all levels. Much of the political map of Africa is a result of the intentional partitioning of land by colonial powers. Preconceptions of order affected the form of American political boundaries and the look of the American landscape. One of the major moulds for the American landscape was the rectangular land survey, That is, the dividing of land into increasingly smaller rectangular counties, townships, sections, half-sections, quarter-sections and so on. This ordinance moulded farmstead location and orientation, farm boundaries, transportation networks, and the shapes and locations of settlements and political boundaries.[26]

Almost any act of planning brings images of order to bear on the landscape. In these cases the preconceived images may be compromised by the need to accommodate to the socio-economic and political realities. Such compromises though need not occur in Utopian thought. Because Utopian conceptions do not have to be practical or even possible, they, more than any other, may reveal the purest forms of images of order at a geographic scale.[27]

Planning, Utopias, conceptions of *terrae incognitae* are all examples of thought in which cognitive maps play a role and in which the spatial distortions and their effects on behaviour can be understood in general without formal hypothesis testing. Unlike the more quantitative studies of cognitive maps, these examples are unsolicited. They come spontaneously from the general process of reflecting about the environment. But these, too, could be analysed in a more quantitative and hypothesis-testing manner, as were the solicited cognitive distances and maps; and more precise relationships about distortions and behaviour could be obtained. Yet we should not be too hasty to do this, for if we rush to describe these images in terms of their parts and relationships to other variables, we may run the risk of overlooking the primary forces which have moulded them. We might not be able to discern the effects that the subjective has had on our conceptions of spatial relations.

In the spontaneous examples of cognitive maps, as well as in the solicited ones, the maps are constructed images using recollection and/or creative imagination. The less the constructions rely on recent or short-term memory, or on memory at all (as in the images of *terrae incognitae* or Utopias) the more likely it is that the images, of either real or imagined places, are moulded by deeply subjective factors and that the images are more closely associated with feelings than with facts. Shapes and patterns symbolise feelings, and these patterns may be used to represent place because of feelings about the place, and not because the place has these shapes and patterns. In general, the associations of spatial patterns with feelings and with place may never be easily related to actual experiences with landscapes. One's sense of a Utopia, for example, may be based on feelings of isolation, stability, order, harmony and justice, which for some may be joined to such spatial images as separation, enclosure, circularity, repetition and uniformity. Elements of past experiences may be involved in these images, but they are moulded by the subjective; and an individual may never have experienced such associations in the environment. Perhaps some images and their subjective meanings are commonly shared, and these more than anything else may affect the way we evaluate and control the environment.[28] To explore in more detail the relationships between spatial configurations and feelings we need to go deeper into the area of the subjective; especially into realms like the visual arts which symbolise feelings in spatial forms.

Spatial patterns and the subjective

In art and also in dream images, the conceptual separation of space from substance is accomplished through the illusion of space which these modes create, and the significance of space is in the emotional import of these illusions. Dreams are examples of unsophisticated thought but we will discuss them here in the last part of the section on the sophisticated–fragmented pattern because they offer the most familiar examples of the idea that shapes and other spatial relations have significance in and of themselves and also because the problems involved in assigning meaning to dream images are closely related to problems of meaning in art. The space–substance axis is seen differently in dreams and in art than it is in science. Whereas dreams do not offer a high degree of conceptual separation of space and substance while art does, we find the separation of space in both to be accomplished through the illusion of space which these modes create, and we find the significance of space in both to be in the emotional import of these illusions.

Dreams

Dreams are predominantly visual symbols of feelings. Understanding dreams requires understanding these visual images. Before Freud, theories of dreams were common which offered fixed meanings to dream images. Like ciphers, these images could be decoded through texts containing their meanings. These texts were often referred to as 'dream books'. The theory of dreams which Freud initiated looked upon such interpretations as overly simplistic. Freud's view held that the dream image could be understood best in terms of the function of dreams. Essential to understanding the dream was realising that the dream was divided into two important parts, the *manifest* and *latent* components. The manifest content of the dream was the part we are aware of dreaming. The latent content (also called 'dream thought') of the dream was the true meaning of the dream. The manifest content in a sense is a disguised form of the latent content, a form which allows some of our hidden emotions to pass the relaxed censorship, in sleep, of our superego. Dream interpretation involves associating images of the manifest dream with their latent meanings.

The assignment of meanings to dream images is never clear and unambiguous. They 'suffer' from an inherent fluidity and vagueness of meaning. This is illustrated in the mechanisms which connect the images of the manifest dream content to the latent dream. One of the most important of such mechanisms is *condensation*. Condensation allows a single dream image or symbol to represent or manifest more than one, and perhaps an unlimited number of, emotional impulses or meanings.

We notice that dreams are condensed because their meanings are not self-evident from the manifest content. They require lengthy elaborations to make sense. The most elementary method of condensing meaning would be through omission, wherein parts of the dream thought are simply not translated or projected into the dream content.[29] Condensation can occur also as a composite image. A person in a dream may appear to combine the features of several other people. The nose may come from *x*, the eyes from *y*, the hair from *z*, and so on.

Another important mechanism by which the dream images obtain their meanings is through *displacement*. The meaning of the dream is only obliquely reflected in its images. It seems centred elsewhere.[30]

In addition to having overdetermined and fluid meanings, dream images do not have exact counterparts to such logical relationships in language as 'if', 'because', 'either', 'or', 'antithesis', 'contradiction'.[31] Yet, logical relations can often be inferred from the spatial position and sequence of dream images. For instance, logical connections are reproduced by simultaneity, in which similar or interconnected things appear close together in the dream. Causal relations such as 'because' may be represented in the dream by having the subordinate clause as a 'prefatory dream and joining the principal clause on to it in the form of the main dream'.[32] It appears, though, that dreams cannot express 'either', 'or', 'antitheses', 'contradictions' or 'no'.

The factors involved in linking symbols to meanings make dream symbols very different from the objective meanings of science and are illustrations of what Langer meant by the differences in meaning between representation and presentation. Warnings abound in psychological literature that the dream symbols are to be understood in terms of the entire dream and in terms of the life of the dreamer. This context dependence, coupled with such mechanisms as condensation and displacement, makes dream images very unlike language; for language contains symbols with more stable meanings than dream images, and rules of grammar which are understood independently of the context. Yet, despite the fluidity of meaning in dream symbols, attempts have been made to assign more or less specific meanings to particular images, shapes and patterns, especially to those which recur in typical dreams. We have just pointed to such attempts in the way dream images reproduce some logical relations. The best known examples of the assignment of rather specific latent meanings to dream images are Freud's discussion of sexual symbols. After reminding us again of the context dependence, *overdetermination* and *fluidity* of dream symbols, Freud nevertheless assigns specific meanings to specific symbols.[33]

What distinguishes masculine from feminine symbols are their spatial properties. The former tends towards the linear and angular, the latter towards the curved and enclosed. It does not matter what the substances are; a phallic symbol could be a sword, a finger, an umbrella, a tree. The important consideration is that all of these elements are linear and pointed. Thus, spatial properties tend to be assigned meaning or significance independently of the substance to which they are related in the dream.

Assignment of significance to shape, independent of the substance, also occurs in Jungian theories of dream imagery. Here we point to archetypal images. These are supposed to be recurrent symbols which humanity shares and which spring from deeper sources than our own personal experiences. They are close to being natural and universal symbols of instincts.[34]

Among the archetypes are the mother, and the forms she takes (the life giver, the all-compassionate, the devourer and the divine harlot), the symbols of the anima and animus, the shadow, and most importantly, the mandala. While the precise form or shape of an archetype varies from dream to dream, it is identifiable as an archetype if it shares resemblance to a general image and the most clearly defined of such images among the archetypes is the mandala, which in Sanskrit means circle. Jung interprets the mandala as a symbol of the self, of the quaternary, of the hero, and so on. Its meaning resides in its shape. 'Roundness,' as Jaffe observes, 'generally symbolizes a natural wholeness . . .'[35] a completion and perfection.

Such views about the meanings of dream images are as close as we can come to the assertion that particular shapes, independent of, or despite their connection to substances, have a significance or meaning. But these assertions even when hedged by reminders of overdetermination, fluidity and context dependence have been looked upon by many as the more dogmatic points in psychoanalytic theory. When they are made too insistently, they then tend to be seen as evidence of a belief in magic and myth. Even if Jung and Freud are correct in their specific assignments of meanings to shapes, we would still need to know the meanings these shapes would have outside the realm of dream images. Would a circle or a linear pattern in the landscape have the same meaning as it has in dreams? To answer this would require that individuals be placed in environments exhibiting these shapes and patterns and that the individuals' reactions to them be observed. Since the contexts in which such shapes and patterns could appear are enormously complex and varied, it seems even less likely that in the assignment of meanings to dream images, we will find specific landscape patterns or shapes which will induce the same feelings, have the same meanings, and produce the same effects in behaviour for all individuals. Thus far we know of no such landscapes.[36] This does not mean that shapes and patterns have no symbolic meaning and do not affect our behaviour. Rather, it means that their meanings and effects are ambiguous and varied. They are context dependent, fluid and overdetermined, perhaps even more so than in dreams.

This problem, on the one hand, of knowing that symbols have meanings, and yet, on the other hand, of not being able to clearly identify their meanings and combine them to form statements with predictable import, is the general problem that Langer described as confronting all scientific analyses of modes of thought which symbolise feelings. Perhaps the clearest examples of this problem are seen in the analysis of art.

Art

Unlike dream images, which are spontaneous, uncontrolled symbols of feeling, art is a conscious attempt to abstract and symbolise feelings. Yet the symbols of art, like dream images, are overdetermined and fluid and depend for their meaning on the whole work of art. Attempts at isolating the elements and symbols, assigning meanings to them, and expecting these meanings and symbols to form building blocks like words in a language so that combinations of them could lead to predictable results, have not met with success for the same reason that dream symbols have not been reducible to lexical clarity or to texts.[37] The symbols of art, like dream images and symbols in all modes which primarily reflect feelings, are fluid and overdetermined. It is even difficult to isolate the elements in art which would have meaning. In a painting, are they the colours, the lines, the figures? Artistic elements are determined by the whole, and they do not correspond to the physical material or the objects which the artist employs. 'Paints are materials, and so are the colors they have in the tube or on the palette; but', according to Langer:

> the colors in a picture are elements, determined by their environment. They are warm or cold, they advance or recede, enhance or soften or dominate other colors; they create tensions and distribute weight in a picture. Colors in a paintbox don't do such things. They are materials, and lie side by side in their actual, undialectical materialism.[38]

Therefore, Langer argues, we·are led to the view that art has

> no basic vocabulary of lines and colors, or elementary tonal structures, or poetic phrases, with conventional emotive meanings, from which complex expressive forms, i.e., works of art, can be composed by rules of manipulation. . . . It is easy enough to produce standard cadences, manufacture hymn tunes according to familiar models and some experiential knowledge of standard alternative resolutions, etc.; but such products are at best mediocre, and their import too slight to show any articulation of feeling. The analysis of spirited, noble or moving work is always retrospective; and furthermore, it is never definitive, nor exhaustive.[39]

Artistic structure, she says, 'is nothing as simple as an arrrangement of given elements by half a dozen, or even a dozen, combinatory operations. The techniques of abstraction and projection are largely derived from the opportunities offered by the material, often on the spur of the moment, in a situation that may never be repeated'.[40] Consequently, the work of art

> is not analyzable in any single set of terms. To comprehend its nature one has to take its author's mingled procedures one by one and consider what each of them contributes to the creation of the single expressive form; this sort of analysis is laborious, but it reveals aspects of feeling projected in the work . . . that are not discoverable by any more systematic means.[41]

Even though the symbols of art cannot be decomposed into the equivalents of a language, we know that the outcome of art has import and we often attempt to describe this import in discursive terms, and to find the elements which contribute to its import. Such attempts are the basis of much of art criticism and in this regard, criticism is a part of social science. It seeks to analyse art in terms of general relationships or apply such general relationships in the analysis of a work of art. As there are as yet few, if any, clear associations in art between elements and meanings, each criticism is subject to dispute and further exegesis. Nevertheless, art criticisms are frequently plausible interpretations because there are some general, though fluid, associations between form and feelings which

can be relied upon to some extent.⁴² These associations are similar to
the ones in dream interpretation. They are strong enough to form the
basis of some plausible assignments of meaning to the subjective. The
plausibility of such assignments is often strengthened by the fact that
associations of feelings and forms are frequently made explicit or
conventional by a culture at a particular period and in a specific style.

The visual arts use physical spatial forms and patterns to symbolise
the entire range of human feelings and emotions. Some of these
feelings which are symbolised by the visual arts, as well as by the non-
visual arts such as literature and poetry, are about space, place and
landscape. These works use either visual or verbal images of spatial
relations to create illusions of things and of space, and, in so doing,
they conceptually separate space from things or substances. The
significance or import of the illusion of space lies in its connection: to
feelings such as vastness or confinement. The work need not,
however, be about a particular place or landscape for it to be about
space. A non-representational work of art can symbolise feelings
about space. Moreover, the import of a work of visual art which
must, by definition, use spatial relations, need not be about space at
all. The portrait of a king, though using spatial relations, may be
about stateliness and not about space.

Often artists attempt to create landscapes or places in order to
draw attention to feelings that can be associated with the spaces or
places represented in the work of art. Place or landscape can be
symbolic of the artistic conception of objectivity or science, or
subjectivity or art, and of their interrelationships. In other words, we
may find in an artistic view of place, an artistic conception of the
objective–subjective axis of our conceptual framework. An aesthetic
expression of the two extremes of this dimension is exemplified in an
excerpt from the fifth book of Wordsworth's *The Prelude*:

> once upon a summer's noon,
> While he was sitting in a rocky cave
> By the sea-side, perusing, as it chanced,
> The famous History of the Errant Knight
> Recorded by Cervantes, these same thoughts
> Came to him; and to height unusual rose
> While listlessly he sate, and having closed
> The Book, had turned his eyes towards the Sea.
> On Poetry and geometric Truth,

The knowledge that endures, upon these two,
And their high privilege of lasting life,
Exempt from all internal injury,
He mused; upon these chiefly: and at length,
His senses yielding to the sultry air,
Sleep seiz'd him, and he pass'd into a dream.
He saw before him an Arabian Waste,
A Desart; and he fancied that himself
Was sitting there in the wide wilderness,
Along, upon the sands. Distress of mind
Was growing in him, when, behold! at once
To his great joy a Man was at his side,
Upon a dromedary, mounted high.
He seem'd an Arab of Bedouin Tribes,
A Lance he bore, and underneath one arm
A Stone; and, in the opposite hand, a Shell
Of a surpassing brightness. Much rejoic'd
The dreaming Man that he should have a Guide
To lead him through the Desart; and he thought,
While questioning himself what this strange freight
Which the Newcomer carried through the Waste
Could mean, the Arab told him that the Stone,
To give it in the language of the Dream,
Was Euclid's Elements; 'and this,' said he,
'This other,' pointing to the Shell, 'this Book
Is something of more worth.' And, at the word,
The Stranger, said my Friend continuing,
Stretch'd forth the Shell towards me, with command
That I should hold it to my ear; I did so,
And heard that instant in an unknown Tongue,
Which yet I understood, articulate sounds,
A loud prophetic blast of harmony
An Ode, in passion utter'd, which foretold
Destruction to the Children of the Earth,
By deluge now at hand. No sooner ceas'd
The Song, but with calm look, the Arab said
That all was true; that it was even so
As had been spoken; and that he himself
Was going then to bury those two Books:
The one that held acquaintance with the stars,

And wedded man to man by purest bond
Of nature, undisturbed by space or time;
Th' other that was a God, yea many Gods,
Had voices more than all the winds, and was
A joy, a consolation, and a hope.
My friend continued, 'strange as it may seem,
I wonder'd not, although I plainly saw
The one to be a Stone, th' other a Shell,
Nor doubted once but that they both were Books,
Having a perfect faith in all that pass'd.
A wish was now ingender'd in my fear
To cleave unto this Man, and I begg'd leave
To share his errand with him. On he pass'd
Not heeding me; I follow'd, and took note
That he look'd often backward with wild look,
Grasping his twofold treasure to his side.
– Upon a Dromedary, Lance in rest,
He rode, I keeping pace with him, and now
I fancied that he was the very Knight
Whose Tale Cervantes tells, yet not the Knight,
But was an Arab of the Desert, too;
Of these was neither, and was both at once.
His countenance, meanwhile, grew more disturb'd.
And looking backwards when he look'd, I saw
A glittering light, and ask'd him whence it came.
'It is,' said he, 'the waters of the deep
Gathering upon us,' quickening then his pace
He left me: I call'd after him aloud;
He heeded not; but with his twofold charge
Beneath his arm, before me full in view
I saw him riding o'er the Desert Sands,
With the fleet waters of the drowning world
In chase of him, whereat I wak'd in terror,
And saw the Sea before me; and the Book,
In which I had been reading, at my side.[43]

In these passages, the poet discusses the relationships between science and art, the objective and the subjective. The dominant symbols in the excerpt are the stone and the shell, and the desert and the sea. The stone is a symbol of geometry, science and logic, of the dry and staid order that comes from scientific understanding. It is held apart from the shell which is a symbol of the imagination and instinct, both of which are closely related to the arts. The desert and the sea also express the dichotomy between science and art. The desert symbolises feelings which are often associated with science and can be seen as the environment of science. It is dry, unchanging, without life. The sea can be seen as the environment of the emotions. It is fluid, unpredictable and unfathomable.

The dreamer was reading *Don Quixote*, a work of art about the Artistic Hero. This work of art, about art, directed his gaze to the sea, a fundament for artistic feeling. The dream takes place in the extreme environment of the desert, a symbol of science, and the dream threatens to end with an inundation by the other extreme, the sea. The dreamer awakens with the work of art beside him, facing with terror the sources of art before him.

Human beings cannot exist for long either in the desert or in the sea. Man needs both the dry and the wet, the land and the sea. To the poet, the arts are more important than the sciences, yet both are the paramount expression of human intellectual effort and must be preserved from the impending deluge. And what is the deluge if not the complete and unbridled immersion in the passions? The extremes of art, or perhaps more accurately, the extremes from which the sources of art spring, are a danger not only to the sciences, but to the arts themselves.

Aesthetic reconciliations of the separation of the objective and the subjective often take the form of a created place or environment in which the opposing forces can be mediated. If the overwhelming environment or threat is the ocean, then an island can provide a haven. If it is the desert, then an oasis can be a sanctuary. In either case, land and water, dry and wet, are combined to form a habitable and comfortable place in which the presence of these elements can coexist and sustain human life. The particular form of the synthesis depends on the artist, on the style and on the age. Often the synthesis itself is depicted with latent or explicit ambiguities and uncertainties, for even in art, the synthesis of the subjective and the objective is only partial and tentative.

Art also depicts the relationship between the sophisticated and the unsophisticated patterns of thought. The sophisticated could be represented by the city with its man-made environments, its artificial cultures and its distance from nature; and the unsophisticated could be represented by the natural elements or wilderness, which is primordial and unreflecting. Man living in cities may lose contact with his origins, and man living in the wilderness may lose his humanity. Again, a synthesis is needed and can be attained in a 'middle landscape', a bucolic setting, or an Arcadia. And once again the place itself may contain contradictions (see Figures 4.2 and 4.3).

FIGURE 4.2 *'The Arcadian Shepherds' (Poussin). There is death even in idyllic Arcadia. In this earlier version of the theme the shepherds seem stunned to find such signs in the landscape. Compare this painting with Poussin's later depiction of the same theme (Figure 4.3).*

Such interpretations of what works of art may mean are in themselves only tentative and partial because they are attempted translations of subjective meanings into discursive forms. These

FIGURE 4.3 'The Arcadian Shepherds' (Poussin). In this later version Poussin has the shepherds in a more meditative mood.

attempts, as we have noted, cannot yet be made with scientific exactitude. Neither can we understand with scientific clarity what elements in a work of art contribute to its import. We can only suggest associations which are on the one hand plausible, but on the other, overdetermined and fluid.

Feelings about shapes, patterns and forms, affect our responses to the landscape and our designs upon it. Most likely the fluidity and overdetermination of meaning found in art and dreams is multiplied when we consider the role of feelings in something as complex as the appraisal and response to the environment at the geographic scale. If we are to understand these relationships scientifically, we must unpack the complex links between feelings and form, and determine clear connections among them within the mould of hypothesis testing outlined in previous chapters. If successful, this course would enrich the social science relationships we have traced and extend the domain of science and the objective into the realm of the subjective.

Given the present state of the social sciences, it can be expected that progress in this direction will be slow and tedious, or not at all, according to those who believe there cannot be a science of human behaviour. Whether there can or cannot be such a science is a point we will not engage, except to say that so far there are no reasons in principle why there cannot be, and to abandon the attempt without offering an alternative mode of understanding human behaviour is to despair of knowing. Meanwhile, we can also pursue the associations of feeling and spatial relations by exploring artistic expressions of place (and we could describe more fully and aesthetically our own feelings). But unless we are more successful in unpacking symbols of feeling in a scientific sense, systematically exploring place and spatial relations in art will not bring us unambiguous systematic understanding of man's relationship with the landscape. Rather, such explorations will edify and result in plausible explanations, but each analysis may be different and always controvertible.

The problems that the social sciences have encountered in their analyses of feeling and art reflect the degree to which the subjective and the objective have been separated and the degree of symbolic specialisation in the sciences and in the arts. Another approach to the understanding of feelings about space, and one which we shall explore at length, is found in those modes of thought which do not conceptually separate the subjective from the objective to the extent that is found in the sciences and the arts. To explore further the subjective meanings of space and their relationships to the scientific perspective we will turn to the modes of thought in the unsophisticated–fused pattern. This is in fact the direction which is pointed to when psychoanalytic theory looks towards early childhood and even less technologically advanced cultures as sources for understanding how dream images and the other psychological symbols are selected to represent feelings. It is the direction suggested when we look for roots or sources of meanings in etymologies. It is the direction to take in the general search for the primordial associations of facts with feelings. The unsophisticated–fused pattern allows us to contract the conceptual surface so that a single mode can encompass the subjective, the objective, space and substance.

Part III
Unsophisticated–fused
Patterns

5
The Child's and the Practical View of Space

In this chapter we are considering the core of the unsophisticated– fused pattern of thought – the evolution of the child's view to that of the practical view of the untutored adult. In these views we do not find a high degree of conceptual separation between subjective and objective and space and substance. What conceptual separation develops depends on the use of symbols. In general, we can say that the growth and development of individuals and species is character- ised by several interrelated processes: the increasing complexity of the organism, its increàsingly complex assessment of its environment, and the increasingly complex interrelationships between the environ- ment and the organism.[1] In the earlier stages of development the organism's parts are not very specialised. There are few centres of organisation and control. The activities of the organism are not well integrated hierarchically and a few or even one action may often involve the entire body, so that the organism cannot attend to more than a limited number of things at a time. In such circumstances, an organism's actions are said to be *global* and *syncretic*.[2] In later stages the actions become more *discrete, differentiated, articulated* and *hierarchically integrated*.[3]

In human intellectual development, the shift from globality and syncretism to differentiation and hierarchical integration is accom- plished by a growing awareness in the individual of self and environment and of their interrelationships. This development proceeds in two stages. The first is a perceptual separation of self from the world. It begins in the first months after birth and is well established by the end of the second year when the child has developed a relatively clear and stable perception of its immediate surroundings and of itself as both an entity or being apart from the rest of the world, and also as one physical object among others in the world. Although from this early stage on, the human being distinguishes self from world, objective and subjective facts are still closely connected and are often intertwined in the process of evaluating self and world. There is a strong tendency in the earlier years, which is still evidenced later on and in other modes of thought, to project one's emotions on to objects of perception. Such animistic projections make the world seem vital and alive. *Animism* is a product primarily of syncretic orientation and is found also in other modes of thought such as myth and magic.[4]

The second stage is the development of conceptual or symbolic thought, and its beginnings overlap with the end of the first stage. Conceptual or symbolic thought is the ability not simply to perceive but to represent oneself and the world through thoughts and symbols. A primary effect of symbols is to drive a wedge between the subjective and the objective, to further differentiate and separate self from world. But there are factors in symbol formation which operate in the other direction, making the separation only partial and tentative. The mechanisms countering the separation, and fusing the subjective and objective, are seen in the earliest stages, where symbols often seem to possess both objective and subjective attributes of the things they represent. The separating functions of symbols are also present at this stage and they become more dominant with time as the symbols become more abstract and specific in their denotations. The symbols seem more conventional and are less apt to be confused with what they represent. Such developments help to further distinguish between and conceptually separate the subjective and objective. Yet the tendency to project emotions on to objects, to animate them, and to view symbolic vehicles as though they possess part of the attributes of their referents, never disappears. Its several manifestations help to break the tendency towards conceptual separation of objective and subjective, and even become dominant in certain expressions and modes of thought like myth and magic.

We shall consider first the primary tendency of symbols to separate self from world which accompanies the shift from the relatively global and undifferentiated view to a relatively differentiated, articulated and hierarchically integrated one; and we will emphasise the objectifying function of this separation. We will consider how the world of objects is seen and conceived in the child's mind and in the practical view of the adult. Then we will consider the ways in which the objective practical view of the child and the untutored adult are interlaced with the subjective.

The objective view of the world

To the neonate, there is no subject and object. 'All that we are fully justified in assuming for the mentality of the newborn child is a blurred state of consciousness in which sensorial and emotional phenomena are inseparably fused.'[5] At the beginning of the child's life, things external to the child have meaning to it only in terms of its sensori–motor involvement with them. By involving the thing in an action, the thing and the action become fused. One of the first reactions of the neonate is to suck on anything that touches its mouth. The things that first come in contact with the mouth are undifferentiated to the child. They are simply things-to-suck. Even later when the child distinguishes from among several objects, the objects have meaning only in terms of their involvement in the infant's elementary sensori–motor activities.[6] In the beginning, awareness itself depends on how things are incorporated into action.

This relationship between awareness of things and eventually of the self, and the assimilation of things into action, becomes more diffused and complex as the child develops. Yet it persists in adult views and in other modes of thought. In early childhood, when it is most intense, it gives rise to feelings in the child that the objects of perception, incorporated into actions, are in themselves active, dynamic, alive. Despite these feelings, normal children develop a relatively stable and objective view of themselves as objects and of other objects in space and time. What is distinctive about the child's view is that the objective is never separated far, or for long, from the subjective. The child is able to shift from one to the other and combine them with far greater facility than the adult, and especially the adult who sees from the sophisticated–fragmented view.

The child's development of a stable perception and conception of objects and space is most thoroughly understood from a Piagetian framework.[7] Piaget describes the parallel development between the organism and its relationship to the environment in terms of two functional invariants, *adaptation* and *organisation*. *Adaptation* itself is composed of two interrelated components – *assimilation* and *accommodation*.[8] *Assimilation* is the process by which the organism incorporates and thereby alters elements in its surroundings. For instance, when we adapt to our environment by eating certain foods, the physical and chemical properties of the food are changed by our digestion. But for assimilation to occur, our bodies must possess the flexibility to accommodate to new situations. To accommodate to food we must be able to open our mouths, to excrete juices, to have our stomachs expand and contract, and so on. Accommodations are the changes one organism needs to undergo so that it may assimilate elements of its environment. Every assimilation requires accommodation and vice versa and both presuppose an organisation. Biologically an adaptation is eventuated when there is a balance between the assimilation and accommodation. This functional balance is more like a dynamic equilibrium. It is always ready to expand to new accommodations and assimilations and is the process which leads to growth. The functional balance, or quasi-equilibrium, is patterned or transformed into the biological organisation of the organism. The organisation is, in a sense, the state of the system, the way in which the equilibrium between accommodation and assimilation are structured. In intellectual development the structures are referred to as *schematas*.

These functional invariants are essential for intellectual development in Piaget's theory. Intelligence can be thought of as organised adaptive acts which involve a balance between accommodation and assimilation. Intellectual development is an hierarchical series of quasi-equilibriums. The elasticity of intelligence is characterised by the fact that new situations can be assimilated to given intellectual structures, but, in turn, the assimilation results in accommodations and thus changes the structure and sets the stage for yet other assimilations.

Assimilation and accommodation result in general structured patterns or schematas. Once schematas are established, they appear over again in slightly modified form as part of higher level mental structures. Development is a series of such equilibrium processes using existing structures to develop new stages of accommodations and assimilations. Take, for instance, the young child of about four to five months who has learned to grasp large objects. The coordination of fingers, hands, eyes and posture are part of the grasping schema. The acquisition of this ability is an accommodation which allows the objects to be assimilated into the environment of the young child. Although the child can grasp larger objects, he must wait until he is about eight months old before he can pick up smaller objects because grasping smaller ones requires a greater degree of coordination. When the child finally comes to grasp smaller objects, he is using, in a modified or more refined form, the earlier grasping schema. Thus, the older schema becomes incorporated into the newer to allow a higher degree of accommodation and assimilation.

Of the many levels of intellectual development, Piaget identifies four major stages, each having numerous and sometimes varying substages.[9] The first major stage is the *sensori-motor*. This stage covers the transition from a purely biological intelligence of the neonate to the beginnings of intelligent behaviour or symbolic behaviour of the two-year-old when a conception of world and self becomes possible. It is by the end of this period that the child has developed a mature perception of physical space and objects. He is able to conceive of his own movements in space and to separate these from the movements of other objects and he understands that he and other objects exist and move in a continuous spatial manifold. He perceives a stable space-time system filled with permanent interacting objects of which he is one.[10]

Piaget emphasises that the conception of space in the child is not developed independently of substances or things; it is thought of as a property of substance.[11]

'Awareness' of space is dependent on the construction of object and self. At the beginning, objects do not exist to the child and we should not expect the child to 'see' what we as adults 'see' or to have the same sense of space as we do. The increasing perception of spatial relations of objects can be generalised mathematically as a transition from topological spatial relations to projective and Euclidean ones. First, the child is aware of propinquity or proximity, then separation, order or spatial sucession, enclosure, and then continuity.[12] The projective and Euclidean properties occur after the coordination of the senses and the establishment of object permanence. Then the child begins to perceive the constancy of size, shape, and the recognition of straight lines and angles which are requisite for projective and Euclidean views.

After completion of the sensori-motor stage, the child begins the three stages of development associated with symbolic operations. Symbolic operations are built upon sensori-motor experience, but they differ from them in four interrelated respects.[13] First, the mental operations at the symbolic level are far more rapid than the sensori-motor and allow for simultaneous considerations of alternatives which would have to be undertaken in a long and temporally ordered series of actions at a sensori–motor level. Second, symbolic activity, unlike sensori–motor activity which is directed to practical needs and satisfaction, can be an end in itself or can lead towards reflective satisfaction. Third, the scope of the child's symbolic functions is not limited to the immediate present surroundings and to physical objects. Fourth, through symbolic functions the child becomes aware of conventional social systems and is able to enter into social relations.

The first stage beyond the sensori–motor is the *preoperational* (two to seven years). This stage is primarily a long transitional period to the concrete operations. It is characterised by logical inconsistencies and lack of regularity in thought. The dominant model for conceptual thought seems to be closely tied with actions. At the beginning of this period the child conceives of objects in a very egocentric way, in terms of remembering or picturing his physical relations with them; and such images of previously manipulated or perceived objects are often uncoordinated and shortlived. Although he has a perception of a spatial manifold, he does not possess a conception of a single spatial continuum, nor does he think that objects in space maintain their shapes and positions independent of his viewpoint. His conception of spatial relations is primarily topological.

In the second stage of symbolic operations, called the *concrete operational* (seven to ten years), the child can coordinate different perspectives and imagine this from another person's viewpoint. He is able, for instance, to imagine the different perspectives what will accompany someone's bodily changes in location and orientation in space, and he notices that space has invariant properties independent of the objects in it. At this stage he is mastering a conception of space which is projective. He now sees that the distance between two objects remains the same when a third object is placed between them, that the winding path between two points is longer than a straight-line path between another two points the same distance apart. This is part of his growing awareness and flexibility of perspectives and his coordination of viewpoints. Projective geometric conceptions of space require such mental coordination of viewpoints. Topological conceptions do not.[14]

The last stage is *formal operations* (eleven to fifteen years). At this level the child is able to think about thought itself and readily to consider possibilities in addition to actualities. At the end of this stage he finally attains a complete and mature conception of space, of a three-dimensional manifold, the spatial properties of which appear stable and Euclidean, and which provide a framework for possible as well as actual events.[15]

While the mature conception finally is of a spatial system, this should not be understood in the sense of a mathematical space described in terms of a geometry. Rather it is a system of space grounded in and derived from substance and their spatial properties and interrelationships. At the end of development, the normal individual may possess a sense of space as a 'container' independent of the things or substances which it contains, but this sense for the untutored adult is not, and cannot be, conceptually explored and removed far from substances, because to do so requires the ability to express systematically its characteristics in the language of geometry, a language which is not available to the untutored adult.[16]

The close association of space and substance, while resulting in the objective practical conception of space, also serves as the focus for the introduction of feelings about space. As we have noted, the cognitive schemata are based on the internalisation of the sensori–motor schemata. This is especially true for the conception of space. According to Piaget:

Spatial concepts are internalized actions, and not merely mental images of external things or events – or even images of the results of actions. Spatial concepts can only effectively predict these results by becoming active themselves, by operating on physical objects, and not simply by evolving memory images of them. To arrange objects mentally is not merely to imagine a series of things already set in order, nor even to imagine the action of arranging them. It means arranging the series, just as positively and actively as if the action were physical, but performing the action internally on symbolic objects.[17]

The role of visual imagery and internalisation of movement may well recede as conceptualisation advances but there will always remain traces of it even in the highest forms of mathematical geometry.[18] What is more, traces of internalisations of movement may be manifested by adults in practically every form of thought.[19]

We attain a practical conception of spatial relationships by internalising what would be our bodily engagements and manipulations of the objects and relationships we are imagining, and by de-emphasising our emotional involvements with and evaluations of these objects. This de-emphasis is reinforced in much of the experimental literature by selecting emotionally neutral objects for the children to manipulate. They are asked to move blocks or vessels, not their teddy bears or security blankets. But the emotional involvement with elements of the environment is not far behind. It often comes to overlay the normal practical view of space. In the aesthetic, mythical–magical and child's view it may in different ways overwhelm the practical view of space, while it recedes and almost vanishes in the scientific mode.

Emotional overtones in the perception and conception of space

One of the major problems of the psychology of visual perception is to account for our awareness of a uniform and stable perceptual field of objects and space while recognising the fact that such a field is constructed from partial and limited views and sensations.[20] Many of the limitations are based on our biological asymmetries. We are upright and we never see more than a part of the visual field at a time. Yet it is precisely such biological oppositions and asymmetries, which are overcome in normal perception, that form the foundations of the intrusion of feelings into the practical view of space. Since these oppositions have a biological basis common to all human beings, there may be emotional elements in spatial views that are common to all men.

The most persistent and evident emotional overlay comes from the oppositions of up–down and front–back. We see what is in front. What is behind is invisible. This direction affects the value we place on 'forward versus backward' orientations. Similarly, we value 'up' differently from 'down'. Our normal erect posture allows us to see from above and is attained by pushing upward against the universal downward pull of gravity. This effort, and the vantage in the visual field that comes from verticality, makes us feel differently about up than down. The different evaluations of front, back, up, down, are feeling tones overlaying visual and conceptual fields. Sometimes they are of little importance, while at other times they may dominate our evaluation of space.

Evidence of these asymmetrical feelings towards space is found in our symbolic systems. In general the front and up are positively evaluated and the down and back are negatively evaluated. Up is heaven, down is hell; up is light, down is darkness. We always look ahead towards success and never turn our backs on it. Such asymmetrical and emotional charges can be projected on to the landscape at all levels. Rooms are assigned fronts and backs and the guest lecturer stands at the front, facing the back.[21] Even the cosmos is given an asymmetry as when heaven is above and hell is below.

Another source of feelings overlaying space comes from the fact that normal conception and perception of space are based on a coordination of the senses, through sight, hearing, smelling, touch, and so on. Variations in the degree to which the senses participate, create different experiences of space. These variations can come about because one or more of the senses is impaired, as in blindness or deafness, which drastically alter the sense of space. Or they can come about when the environment provides over, under, or no stimulation for one or more of the senses as in the case of the snowy white environment of the polar regions which drastically reduces a novice's ability to perceive distances and directions and to find his way. Long exposure to such environments makes a person rely on other senses to compensate and to regain a full practical perception.[22]

Cognition of spatial relations is altered by the affective qualities of the objects in space. To someone living in the midwest, New York City may seem closer than it actually is because of the warm feelings the person may have towards the city, or to people in it. On the other hand, different emotional evaluations of things can occur simply because of their relative locations. This stems from the fact that we may expect or prefer things to be in certain places or spatial arrangements and we may think less of them when they are not. For instance, in normal face-to-face conversation, we may prefer to have a clear view of the face of the person with whom we are speaking, while not caring to be bathed in that person's breath. We would therefore want the person to be close enough to see but not so close that we will smell him. A person who stands either closer or farther than this will make us uncomfortable and earn our resentment. After much experience we tend to develop a sense of zones around our bodies which define the proper or comfortable distances for certain activities. These personal spaces may be culturally accepted and standardised, and if we know the personal distances of a culture, we can use them symbolically. We can show belligerence by standing too close, or disdain by standing too far. Such differences in feeling about things and people, because of their relative locations, create emotional overlays to physical space. While such zones can be estimated in general, they differ considerably from culture to culture and from individual to individual.[23]

Properties of space can seem to have emotional value especially when objects of intense emotional import appear in specific places and in regular spatial patterns, shapes and forms. People exposed to such environments may associate the emotions with the shapes and physical spatial positions of the objects. The sun is perhaps the single most important source of wonder. Given its spatial regularity and importance, many cultures associate roundness with cosmic power, and the direction of the sun's origin with positively valued things and emotions.

There are awe-inspiring elements of the landscape which are not common to every environment and which impart emotional value to places. In some cultures, mountains are sacred and express the positive associations with the vertical (for example, in Indian mythology, Mount Merew the sacred mountain was thought to be at the centre of the world). In other cultures rivers are sacred and express and organise the evaluation of space along a horizontal (the Nile river and the sun were the dominant physical features organising the Egyptian's evaluation of space). In other cultures, other features of the landscape, which to the outsider may have no significance, have emotional import and anchor the emotions to places. Such places become a part of the mythology of a culture. They become holy places, places which are believed to have been created and moulded by gods and spirits.[24]

Bodily asymmetry, the combinations of the senses, the differential evaluation of objects in the environment, all contribute to the overlays of emotion on to space and spatial properties. While these overlays occur naturally and universally as a result of such fundamentally shared factors as biological and perceptual asymmetries, their specific contents and their effects on behaviour differ from individual to individual and culture to culture. From the scientific viewpoint, their meanings are not as predictable as our feelings would have us believe. People from similar environments have different cosmologies, different personal distances, different holy places and different symbolic systems. Although the sun is almost always a symbol of life, and the east, as the place of the sun, is usually thought of positively, the precise attributes associated with the sun and cardinal directions differ from culture to culture.

The sun is the primary source of light and heat and it is always round. Therefore, it seems natural to associate the circle with such things as light, power and the things they connote. Such associations are strong and make possible the belief that, in a shape like the circle, we are glimpsing a *universal* and *natural* symbol whose objective and subjective meanings are at once naturally and universally understood and shared and which seem to inhere in the spatial properties of the symbol. But the association is compelling only within a certain frame of mind and then perhaps only sometimes. And even on these occasions the subjective meanings seem to be extremely fluid and so broad as to often be contradictory. The ideas of 'power' and the 'circle' may connote (what from the logical perspective are) contradictory feelings of boundedness and unboundedness, of the finite and the infinite. In addition, the circle assumes other and numerous meanings which are clearly conventionally defined.

Similarly, the place of the sun and other cardinal directions have different meanings in different cultures. The ancient Chinese associated east with green, anger and wood; west with white, sorrow and metal; north with black, fear and water; south with red, joy and fire.[25] The Zuni associated east with earth, seeds, frost and white; west with water, spring, and blue; north with wind, winter and yellow; and south with fire, summer and red; and the Oglala Sioux associated east with red, light, morning star and wisdom; west with black, thunder brings rain; north with white, and great, white, cleansing wind; and south with yellow, summer and growth.[26]

We find examples of general agreement and yet extremely wide divergence of meaning in other fundamental spatial orientations and relationships such as right–left and up–down. While up is usually valued positively and down negatively, we can find exceptions to this association in our own language, as when we speak of something on top or on the surface as being superficial, and something deep down or beneath as being fundamental and profound.

There are then general tendencies to have what appear from certain perspectives, especially those of myth and magic and some areas of aesthetics, to be natural and universal meanings for some spatial relations, meanings which appear to inhere in the spatial relations. Yet when these meanings are examined outside these views, and from the logic of science, they are not universally shared. This does not deny that from particular perspectives or modes, such symbols can be felt to be universal and natural; and that such feelings, having bases in the biologically common mechanisms which guide the projections of our emotions on to space, leave us with the capacity to understand how such feelings can exist and even with the expectation that natural and universal symbols will be found objectively to exist. We know that symbols can appear to fuse both feelings and fact, but we also know that we cannot yet predict which symbols will do it, nor the outcome or details of their associations.

We have turned to the unsophisticated–fused views, starting with the child's view and the normal practical view of the adult, in order to observe the most basic and general conditions in which the subjective and objective become closely related, and in which symbols seem to contain attributes of their referents. While the child's view and the normal practical view indeed do contract the subjective–objective axes, the variety of ways in which the subjective and objective are combined and symbolised is still not based on clear rules of association. In unpacking and analysing the fusion taking place in these views, we once again encouter the problems of determining clear and unambiguous meanings for subjective terms and states. The same problems we encountered in the interpretation of art are involved in unpacking meanings in the unsophisticated–fused view. We may feel that symbols can have natural and universal meanings, and yet we may be unable objectively to characterise these meanings. This is a paradox most sharply drawn in the study of two psychological processes, synaesthesia and physiognomic seeing. These two processes would appear to be central to forming symbols about space which seem naturally to express both subjective and objective meaning. Moreover, both processes have been said to occur more commonly in children than in adults.

Synaesthesia is the process whereby a secondary sensation in one field is produced by a primary stimulus in another.[27] Synaesthesia can occur between any pair of perceptual modes and can involve almost any sensations and images within them. A sound may produce the sensation of seeing a colour, a shape may produce the sensation of sound, an odour may produce a feeling of warmth, a number may produce a sensation of colour and so on. Experiencing such associations tends further to animate the world by integrating sensations. Furthermore, if one member of the pair is a symbol for the other, as when a colour may stand for a sound, the symbolic vehicle appears to share naturally the properties of the things it represents. Synaesthetic experiences can be divided into two types: a strong, genuine or sensory synaesthesia, in which the secondary sensations are involuntary, and a weaker or associative one in which the secondary sensations are more like a conceptual association. These two may in fact form the extremes of a continuum and may even be interrelated.[28] On the sensory side, some subjects 'see' or 'hear' the secondary sensation as though it were out there, as though it were a real thing, affixed to the primary stimulus. For others the experience may be more diffuse, the primary stimulus calling to mind the secondary sensations as when we say a colour is loud.[29] For some, associations are clearly extrinsic and explainable by the correlation of perceived experience, memory and unconscious associations, as when a particular sound evokes the image of a particular concert hall. But even such extrinsic associations may be forged by some intrinsic similarities between the sensations, and hence, there again is the possibility that certain stimuli are naturally associated with others.[30]

Although in Western cultures intrinsic synaesthesia seems to be experienced by only a few adults, especially artists, either extrinsic synaesthesia or an appreciation of synaesthetic associations is experienced by most individuals when it comes to aesthetic expressions of meaning. A (near) universal connection of some sort naturally comes to mind to explain the power and appeal of sensuous associations and metaphors in literature, and such associations are important components of literary expressions. In a sentence such as

> I am engulfed, and drown deliciously.
> Soft music like a perfume, and sweet light
> Golden with audible odours exquisite,
> Swathe me with cerements for eternity.[31]

they are the dominant form.

Sensuous metaphors and appeals to synaesthetic associations form the basis of one of the most celebrated passages in the English language. From *Twelfth Night* we have

> If music be the food of love, play on!
> Give me excess of it, that, surfeiting,
> The appetite may sicken, and so die.
> That strain again! It had a dying fall;

> O, It came o'er my ear like the sweet sound
> That breathes upon a bank of violets,
> Stealing and giving odour.[32]

But elsewhere Shakespeare satirised such associations and does so by bringing together symbols that are not synaesthetic: 'I see a voice! Now will I to the chink,/to spy and I can hear my Thisby's face.'[33]

There have been attempts to depict aesthetically the sounds of musical compositions in the medium of shapes and colours, as in the film *Fantasia*. The sounds of poetry have been rendered into shapes and geometric patterns – as in Marjory Pratt's depiction of a Shakespearean sonnet, (Figure 5.1) – and there have even been attempts at mechanising particular associations of sound and colour in colour organs.[34]

Since brass, .nor stone, nor earth, nor boundless sea,
But sad mortality o'er-sways their power,
How with this rage shall beauty hold a plea
Whose action is no stronger than a flower?
O, how shall summer's honey breath hold out
Against the wreckful siege of battering days,
When rocks impregnable are not so stout,
Nor gates of steel so strong, but Time decays?
O fearful meditation! where, alack,
Shall Time's best jewel from Time's chest lie hid?
Or what strong hand can hold his swift foot back?
Or who his spoil of beauty can forbid?
　　O, none, unless this miracle have might,
　　That in black ink my love may still shine bright.

FIGURE 5.1　*Syllabic design of Shakespearean sonnet*

Thus far we do not have general rules of association of the senses which would make reliable that translations from one sensory mode into another would have aesthetic import. The success of such translations is as unpredictable as are the creations of any work of art. If we knew in general the way people associate these senses, we could have a powerful tool for explaining and predicting the emotional imports of shapes, forms and patterns, and a tool to help us analyse how a symbolic vehicle in one sensory mode can be a natural and universal symbol of sensations in other sensory modes. Although there are as yet no general synaesthetic relationships which can be used in combination to yield predictable results, there are tantalising suggestions of relationships among the senses. For instance, it seems that synaesthetic associations tend to make correspondences between the energy levels of sensations,[35] that the sensuous metaphors generally transfer senses from the lower to the higher ends of the sensorium, that is, from tasting and touching to sound and light,[36] that red gives the impression of being a warm, heavy, advancing colour, giving the sensation of 'reaching out', while blue gives the impression of being light, cooler and receding.[37]

Such associations point to general tendencies, but they are overdetermined and fluid like the associations in dream images and art. The evidence thus far suggests that the associations do not hold in different contexts and cannot, as Langer points out, 'be used in combination to produce predictable results'.[38] Despite these problems there has always been the hope of finding a general translation of one sense mode into another, and thus of identifying universal symbols that are able to manifest meaning in more than one mode. We often personally make such translations in our own private synaesthesias: some of these associations are culturally recognised and become part of a symbolic system; and some are employed by great works of art. Moreover, philosophies of art and magical systems have been built on the premise that there exist natural, universal symbols which would have meaning in more than one sensory mode.

Physiognomic seeing is the perception of objects and of humans as though they possessed emotions and feelings. As with the synaesthetic experiences, physiognomic seeing leads to the animation of objects and to the tendency to see symbolic vehicles as naturally expressing emotion or feeling. Physiognomic seeing, even more than synaesthesia, may be the vehicle to disclose how properties of space and spatial relations are associated with feeling.[39]

For the child, physiognomic seeing occurs normally and frequently. The child does not have sufficient distance between himself and the objects to shift perspectives and see that things do not literally possess these feelings he imparts to them. What we take for a tilted cup, a child may see as a 'poor, tired cup', what we see as the number 4, a child may see as soft, or the number 5 as mean and cross, or a simple towel hood as a 'cruel thing'.[40]

Simply thinking about objects in a syncretic fashion makes them appear to embody sensations or emotions. As adults we may experience such sensations when, in a specific attentive attitude, elements in the landscape seem to possess emotions. A landscape may be forlorn, or the bent willow is sad and weeping. In a sophisticated view we are aware that this is only an attitude; the willow is actually not weeping nor is the landscape solemn, although from the physiognomic attitude they may appear that way. In physiognomic perception we are projecting our emotions on to these objects and we usually recognise this by saying that we mean 'forlorn landscape' metaphorically. We may also create physiognomic experiences by using shapes and patterns to evoke emotions. The clearest examples of such creations are animated letters, which give the appearance of possessing such emotions as happiness or sadness. Even though, as adults, we are aware that inanimate objects do not possess emotions, physiognomic perception, when it occurs, is convincing.

Experiments have been performed on adults to create a physiognomic perception for even the most unexceptional objects or symbolic vehicles. For instance, if words are to be viewed physiognomically, subjects may be asked to 'alter their everyday manner of dealing with . . . words, so that these linguistic forms tangibly–visibly and/or auditorily–portrayed their meanings'.[41] The results of such experiments show that often these suggestions are taken up; the subjects do in fact see printed words as possessing physiognomic qualities. For example, for the German word wood, *Holz*, in one experiment, the response was 'something crude, raw, uncouth. One gets stuck at its splinters if one moves over the word with one's eyes. This quality seems to be (visually and articulatory–auditorily) centred particularly on the o and the z', whereas for the German word *wolle* (wool), which seems to look soft, the 'Softness lies particularly in the w and o'.[42]

The clearest experiments on physiognomic perception of spatial relations have dealt with responses to non-conventional linear patterns (that is, lines such as ⌒ ⌐ ⌒ ∿∿ etc.). Such experiments often have subjects graphically express objects or feelings by such lines (as when we are asked to draw happiness); match a given set of lines with a given set of objects or feelings (as when we are asked which line, ⎍⎍⎍ or ∿∿∿ , more represents gold); or 'open endedly' describe what feelings or objects certain line patterns connote. When the lines are seen physiognomically by the subjects, the physical properties of the line are completely overlain with emotional import.[43]

Physiognomic experiments reveal idiosyncratic and variable assignments of meanings to lines. The same or similar geometric forms may be seen physiognomically quite differently, while the same or similar feelings or concepts may be expressed by quite different geometric linear forms.[44] For instance, two subjects produced essentially the same geometric pattern, one for 'blue' ($\vee$) and one for 'stone' ($\vee$). Yet, each of the line patterns was seen differently physiognomically. 'The line pattern for "blue" was taken as an "emanation from inside, outward and downward–more or less as blue light emanates down from above"', and 'the pattern for "stone" was intended primarily as an expression of "pointed, penetrating features"'. On the other hand, the pattern (⌐⌐) was produced to represent 'longing' for one subject while ( ) was produced to represent 'longing' by another.[45] But there are also degrees of consensus among respondents, especially in experiments which limit the number of lines and meanings to be compared. For example, the elements gold, silver and iron were to be matched with the lines (a) ⌐⎍⌐⎍ , (b) ∧∧∧∧∧∧ and (c) ∧∧∧∧∧. Given these limitations, the majority of respondents paired iron with (a), silver with (b) and gold with (c).[46]

The occurrence of both a degree of consensus in the meanings of line patterns and individuality of meaning indicates that the same mechanism of assigning meaning in dream images and art symbols are at work for physiognomic seeing. Heinz Werner identified the equivalents of fluidity and overdetermination of meanings in physiognomic seeing and termed them *plurisignification* and *polysemy*. *Plurisignification* is the situation in which one symbolic vehicle when seen physiognomically can become several symbols and *polysemy* is the situation in which one symbol represents several meanings.[47] These processes, of course, are very similar to the mechanisms which Freud identified in dream thought.

The paradox of consensus on the one hand, and individuality of meaning, plurisignificance, and polysemy on the other is again difficult to explain for it directly addresses the problem of rendering the subjective into objective form. Only a few general relationships emerged and these are about the constraints rather than about the intrinsic meaning of line patterns. Consensus depends, in part, upon restricting the alternative meanings so that there is an agreed upon universe of discourse. Instead of having individuals supply any and all meaning that they associate with a line, the researcher restricts the choice by supplying a few alternatives from which the subjects must choose. Consensus seems also to be affected by the degree of differentiation among the referents. It is more easily arrived at when the alternative meanings for a line are discussed in extremes, like love and hate, rather than in smaller degrees, like love and like. Consensus is also affected by the degree of geometric differentiation among symbols. Sharp lines versus curved lines may create a greater degree of consensus (assuming the above) than two patterns of sharp lines.[48]

Factors such as bodily asymmetry, the coordination of senses in the perception of space, the different evaluations of objects in space, synaesthesia and physiognomic seeing all contribute to the associations of spatial properties with feelings. In certain states they may so transform the normal perception of objects and space as to present an active, animated world in which places, distances, directions, shapes and paths are dynamic and affective. Such states occur especially in the unsophisticated-fused views when there is a short distance between the subject and object. Such perceptions are experienced by adults in the western context but perhaps with less intensity, belief and duration. Nevertheless, they occur and when they do, the association of feeling and spatial properties seems immediate, certain and natural. Yet the associations are fluid and overdetermined, and because of this, we know of no stable and general relationships between feelings and spatial properties in the environment which could be relied on to enter into combination with other such relationships to form new and predictable ones.

Space in language

Symbols are required to conceive of a stable spatial manifold populated by objects. The symbols we have considered thus far are often private and not very abstract. The most characteristic symbolic form which allows for intellectual flexibility and abstraction are natural languages; and in language we again find the coexistence of clear objective meanings with powerful, immediate, yet overdetermined and fluid subjective meanings. These two facts play a dialectical role in language development.

Early symbolic use in general occurs within a syncretic framework.

The continuing development of symbol use tends to diminish syncretism by prying subjective and objective apart; but, the symbolic vehicles especially in the earlier stages of development are felt to possess both subjective and objective properties of their referents. In · this regard, they tend to reconnect the subject and object and may sometimes even lead to the confusion of symbols with referents.

The conditions by which word symbols seem sensuously to reproduce or characterise inner and outer feelings have been termed *symbol transparency* or *symbol realism*.[49] This reproduction makes the symbols appear to be the 'natural' representatives of their referents and even part of their referents. In examining symbol realism in language we shall consider especially how feelings about space are sensuously embodied by linguistic vehicles and thus absorbed by language; and in turn, how language further expands and differentiates the meaning of space and projects new feelings and abstractions on to it.

Symbol realism, like synaesthesia, has been an important element in aesthetic expression, and perhaps we are most familiar with it in literature and poetry. It is difficult to find examples in poetry where it is not present, and in some cases it dominates the poem. In Browning's *Pied Piper*, the sounds definitely reproduce the sense.

> And the muttering grew to a grumbling;
> And the grumbling to a mighty rumbling;
> And out of the houses the rats came tumbling.[50]

So too, in Milton's *Paradise Lost*:

> The Serpent subtlest beast of all the field[51]

he hears,

> On all sides, from unnumerable tongues
> A dismal universal hiss, the sound
> Of public scorn.[52]

To some, symbol realism is an essential part of poetic expression:

> 'Tis not enough no harshness gives offence,
> The sound must be an echo to the sense.
> Soft is the strain when zephyr gently blows,

And the smooth stream in smoother numbers flows;
But when loud surges lash the sounding shore,
The hoarse rough verse should like the torrent roar,
When Ajax strives some rock's vast weight to throw,
The line too labours, and the words move slow;
Not so when swift Camilla scours the plain,
Flies o'er th' unbending corn, and skims along the main.[53]

While symbol realism is part of the extraordinarily sophisticated modes of expression in art, it has also been thought to be the first stage in the development of language both for individuals and for mankind.[54] The first sounds of the child, the babbling, chortling and crying, are expressions of emotions which in their very production exhibit expressive quality. Children love to babble for the sake of sensuously experiencing the sounds. This sensuous involvement in sounds creates a dynamic shifting between sounds and feelings. Language uses sounds as symbols, and in early linguistic acquisition there is a strong link between the sound and its objective and subjective referents. Children learn a developed language which contains symbols whose meanings are already established. When learning language, children distort words, make up new ones, combine words, alter grammar, and these alterations contain evidence that children often conceive of linguistic symbols as though they were transparent.

Among the most characteristic features of children's sounds is their attempt to imitate sonically and replicate the sounds of their surroundings, to form an onomatopoetic rendering. But sonic imitation may not only be of sounds, but of the synaesthetic and physiognomic factors associated with the objects and symbolic vehicles in the environment. It is in the sonic renderings of physiognomic and synaesthetic sensations that attempts are found to render in sound what is felt about spatial relations. For instance, a very young child may represent something small by using short vowel sounds, while using long vowel sounds to represent large objects; or the vowel 'i' may refer to something high or distant, while 'u' may refer to something low.[55] Intonation may help express size, as when a child uses a sound with a high, thin, short intonation for pebbles while using a sound with a low, long intonation for rock; or the rolling of the tongue may represent the process of rolling or gliding.[56] In a language contrived by two English children, the word *bal* meant 'place', 'but the bigger the place the longer the vowel was made, so that with three different quantities it meant "village", "town" and "city" respectively. The word for go was *dudu*, the greater the speed of the going, the more quickly the word was used. . . . '[57]

Feelings about space are incorporated and elaborated in language in other ways than through sonic renderings. The emotional association derived from the asymmetry of the body and the force of gravity affects the order of word acquisition. The way in which the effective qualities of space are transmitted is in the *markedness* of terms. For instance, in the couplet 'long–short' the first term is the basis of measurement. 'How long is it?' can be answered in terms of a single measure, that is, ten feet or ten miles. The measure is called the length. On the other hand, the shortness is always relative to the length. We do not say it is ten feet short. In this respect, shortness is derivative of the length and it is a more complex measure than the length. This means it is *marked*. In general, the more complex term is said to be marked with respect to the less complex measure and usually those terms which refer to the positive bodily orientations, such as 'up' and 'front', are unmarked. The second terms in the following couplets are marked: long–short, far–near, tall–short, deep–shallow, wide–narrow and thick–thin. The way in which markings and affective qualities of space influence behaviour in language is that in experiments in word comprehension among children it was found that the unmarked part of the couplet was comprehended earlier than was the marked part. What is more, these values of spatial relations affect adults' use and understanding of language. It has been shown that direction with positively valued terms is comprehended more easily and quickly by adults than with negatively valued ones.[58]

Having words mimic the qualities of space, through onomato-poetic and physiognomic–synaesthetic representations, tends to decrease the distance between symbols and their referents. This close distancing allows the word representing an action to be seen as a part of the action. 'Saying "mama" when the mother appears need not be psychologically different from hugging her when she holds the child in her lap; shouting "mine" can merely be an instrumental act for keeping an object, just as one would cling tightly to it.'[59] Lengthy, rambling monologues often accompany the child's action with no apparent purpose, other than to render the activity in sentences.[60] By physiognomically representing the world as seen through normal perception, words for children become 'much nearer to action and movement than for us'.[61] They become at times objects themselves, possessing the power to affect other things. 'When the distance between two points has to be traversed, a man can actually walk it with his legs, but he can also stand still and shout: "on, on! . . ." like an opera singer.'[62] The general belief that words embody the essence of things and can be used to affect other things is the basis of word magic.

Words may appear to be so real that they can literally be projected into reality on to the objects they represent. Until the age of about five or six, the child believes the name of an object is inside or is attached to the object. Moreover, the child believes he can recognise the names by looking at the object.[63]

As the child grows older, the word may no longer be seen as adhering to the object it represents but it is still real enough to be thought of as a material thing somewhere in space. Only when the child finally uses symbols maturely are they understood as conventions.

The 'outward' projection of transparent symbols extends to spatial relations. Once feelings about spatial relations are absorbed into the structure of language, these spatial terms and relationships can form a structure for newer words and concepts. Words representing spatial relations become anchors for newer words, and hence the referents of these newer words may be projected on to space (even when they do not belong there). When a child is asked to do something he does not understand, instead of saying 'what?' he may say 'where', and when the child learns about an abstract system such as time, the temporal concepts are understood by their associations with spatial terms. We say the future is 'ahead' of us, and they arrived 'before' us. 'Ahead' and 'before' are spatial terms used metaphorically in a temporal context. What is more, the affective qualities of space due to bodily asymmetry are carried through in such metaphorical usages and influence the order in which such expressions are first comprehended.[64]

The shift from transparent to conventional symbols has been argued to characterise the development of language in general. Unlike the analysis of children's language, which can rely on contemporary observations and experimentations, the reconstruction of pre-historic languages relies on very incomplete evidence and on the shaky contention that ethno-linguistic observations from extant primitive societies are about tongues which have not evolved much from the Stone Age and can therefore serve as examples of the earliest forms of language. Any conclusions from such evidence should be examined with a great deal of scepticism. Indeed, one of the more popular current generalisations about language is that no contemporary language is more primitive than another.[65] Nevertheless, there are some languages, whether or not they are primitive, which do sonically render spatial relations in more detail than do other languages, including our own.[66]

Examples of language transparency, of course, can be found in all languages regardless of their stages of development and when it occurs it tends to reify language and invest it with power. The depiction of spatial relations in language also makes spatial expressions an anchor on which to fasten more abstract relationships. Abstract concepts seem to become intelligible as they are placed in a spatial context.

Attempts have been made to identify universal relationships between sound and space. Humboldt pointed to the group 'st' as 'regularly designat [ing] the impression of the enduring and the stable, the sound "l" that of the melting and fluid, the sound "v" the impression of uneven, vacillating motion'.[67] 'A', 'o' and 'u' have been said to designate the greater and 'e' and 'i' the lesser distances and Sapir pointed to the tendency for 'i' to connote a smaller volume and for 'a' to connote a greater volume.[68]

These are tendencies of sounds and words to reproduce sensations, for them to be 'an echo to the sense'. But while the transparency of symbols can be intensely felt, it is hard to find general associations of symbolic vehicles with referents. Such associations are fraught with exceptions and vagueries and do not lend themselves to predictable combinations. The babblings of children, the onomatopoetic rendering of sound, differ from individual to individual and culture to culture in such a way that the elemental units have not been isolated. The crowing sound of a rooster in English is cock-a-doodle-doo, in German it is kikeriki, and in French it is coquerico. The feelings about sound which may hold in isolation do not hold in combination with other sounds and in combination with sounds and words that have conventional meaning. Again, we confront the problem of recognising that symbols naturally reflect feeling and yet being unable clearly and unambiguously to isolate and define the relationships.

6
Myth and Magic

The mythical—magical mode elaborates several characteristics of the childlike and practical views – for example, the use of symbols as though they were natural and universal. Because of this elaboration, the mode occupies a greater area on the conceptual surface than do the child's and the practical views.

Many man-made landscape features owe their spatial forms to our beliefs in myth and magic. To understand these features we must understand the meaning of the mythical—magical mode of perceiving space. Although our interests are in the effect of this mode on the landscape, the richest evidence of mythical—magical meaning is not found on the ground, but rather in symbolic forms such as tales, rituals and incantations. We must direct our attention to these forms of expression before exploring the shapes and patterns on the ground that result from the application of this mode. But first it is important to repeat that by myth we do not mean simply the belief in mistaken ideas as in 'it is a myth that the Sun revolves about the Earth'. Rather, by myth we mean such symbolic forms of expressions as are found in formal myths and mythologies, and which have a structure similar to magical rituals and expressions.[1] It is the unusual structure of mythical—magical symbols that creates the distinctive meaning of space in this mode and that holds the key to interpreting mythical—magical views of place and landscape.

The fantastic

To the Western 'objective' viewpoint, the extremely varied forms of myth and magic all possess a quality of the fantastic. Despite the enormous variation in details and concerns, myth and magic present a world of fluidity and mutability which flouts our scientifically-based concepts of causality. It is common, indeed characteristic, of this mode that anything can change into anything else: people into animals, animals into plants, plants into inanimate objects, and these into animate ones. Things can suddenly appear from nowhere and equally suddenly dissolve into nothing, and actions can occur at a distance. 'By a sudden metamorphosis everything may be turned into everything. If there is any characteristic and outstanding feature of the mythical world, any law by which it is governed, it is,' according to Cassirer, 'this law of metamorphosis.'[2]

Fantastic metamorphoses are evident in virtually every mythical—magical tale. We will very briefly describe the action in some of these tales simply to give a sense of the range of such metamorphoses. For example, there is the belief of the Karamundi of Australia that rain can be brought down by the following procedures: allow blood from a vein in the arm to drop into a piece of hollow bark, to this add gypsum and hair and grind to a thick paste, then place the mixture between two pieces of bark which are put under water so that it will dissolve and then rain will follow.[3] As another example, there is the Tsmishian Indian Tale of Asdiwal. Some of the highlights of one version of this tale are that Asdiwal was fathered by a being who could take the form of a bird. During one of Asdiwal's many adventures, he captures a white bear and follows it into the heavens where it turns into a beautiful woman, the daughter of the sun, whom he marries. He dies and is brought back to life again. He later is transformed for a while into a bird, visits with seals under the sea, and finally he turns into stone.[4] The fantastic is part of a version of a Borroro Indian myth in which a woman who was raped by a member of her moiety is killed by her husband. Their son, who saw his mother raped, turns into a bird and drops excrement on his father's shoulder. The excrement then turns into a tree.[5] More familiar to Westerners, but equally fantastic, are tales from western religions. Take, for example, the Biblical tales that the Earth was created in six days by the force of God's words, that Moses' staff turned into a serpent, or that Christ, the son of God, was killed and resurrected.

Attempting to discover why myths are created and what they mean has occasioned numerous theories; many of which address different aspects of the characters and functions of myth. None so far has deservedly been heralded as the only and most comprehensive one; but among them is an extremely influential one which can be expanded upon more than any other to examine the role of space and causality in myth and magic. We shall refer to this theory as Lévy-Bruhl's, even though there have been several other people who have made major contributions to it. We will draw from and expand upon his theory in our analysis of space in the mythical–magical mode.[6]

According to this view, the fantastic qualities of myth and magic are a consequence of a different logic, a 'pre-logical' form of thought which falls within the unsophisticated–fused pattern of our framework. Myth and magic can occur unselfconsciously when individuals are in extremely close, intense and emotionally heightened association with objects of nature. In Lévy-Bruhl's terms, this situation creates mythical participation, and in our vocabulary, this situation occasions the confusion or conflation of symbol with referent which seems to us to be the pre-logical characteristic of participation and of myth and magic in general. The fantastic is a natural consequence of this confusion.[7]

The vitality and persistence of myth and magic are due in large measure to the frequency and spontaneity with which such heightened emotional states occur, and to the fact that these feelings are expressed in an unsophisticated form. The unsophisticated form of expression disguises the symbolic nature of myth and magic. To those who believe in it, the tale or ritual is not figurative or metaphorical, it is literal.

This view offers the most direct explanation of why so many people in so many cultures employ this mode, defend it and act as though they believe in it. Because myth and magic are an extremely noticeable part of the primitive world view, this mode and Lévy-Bruhl's interpretation of it have been used to characterise the primitive world view and primitive thought in general. Extending pre-logical thought to cover the primitive mind has drawn a great deal of criticism. Opponents contend that the use by primitive people of myth and magic does not seem to interfere with their normal practical affairs, wherein they are capable of distinguishing symbols from referents. It is pointed out, for instance, that although a primitive hunter says prayers or performs rituals over the tracks of a wild animal he is hunting, at the moment of killing the beast, the hunter knows that the tracks are symbols and not the animal itself. He shoots the arrow at the animal and not at its tracks. The hunter may be conflating symbol with referent when he prays over the tracks of animals, but not when he shoots his arrow.

In other words, the conflation may be limited to certain activities only and not extend into all the rituals of daily living.[8] Our own society has innumerable examples of myths and rituals which are confined only to specific activities. Mythical—magical consciousness, as Lévy-Bruhl in his later writings suggests, can exist side by side with other expressions and views of the world.[9] Primitive society, though, has a different mixture of modes than does modern Western society, and myth may enter more of its realms. It remains to be determined why so many facets of their lives are infused with rituals and why these rituals are so tenaciously followed.

Analysing myth and magic as a pre-logical system of expression does not help us decipher the particular meaning a myth and a magical ritual might have, nor does it reveal common structural elements and meanings within this mode. In fact, this analysis does lead us to anticipate that the meanings of mythical—magical symbols will be highly fluid and overdetermined, just as are those of other unsophisticated symbolic expressions. It is this implication of Lévy-Bruhl's theory which seems to us to be in marked contrast to the Structuralist interpretations of myth and magic. The Structuralist view, as developed especially by Lévi-Strauss, has been heralded as a general approach to the meaning of mythical symbols through which common structures in mythical thought can be revealed.

Structuralism is a complex and controversial form of analysis. One of its contentions is that myths are structured in terms of binary oppositions which express important dilemmas and antitheses confronting communities. The subject matter or content of a myth is selected to express these fundamental oppositions and is secondary to them. It is in the structure and pattern of these oppositions that we find common meanings in myth. A myth's structure in a sense transcends its content, revealing the inner workings of the mind.[10]

The primary reason Lévi-Strauss' views about myth appear to be contrary to Lévy-Bruhl's is that Lévi-Strauss contends that the inner meanings and associations of myths follow something like a conventional logic.[11] The same symbolism may have several meanings, but each meaning is fixed within a network of oppositions at a given level of interpretation. How Lévi-Strauss demonstrates this is too complex to discuss here. Critics have asserted that Lévi-Strauss' penetrating and insightful analyses of particular myths may not be dependent on a structuralist approach. They have argued that the use of binary oppositions is arbitrary. Myths may have more complex relationships and oppositions than this, and the binary structures do not capture the rich nuances and fluid meanings which are characteristic of myths.[12] Despite these criticisms, if Lévi-Strauss' views are correct, they do not necessarily contradict Lévy-Bruhl's thesis because the two can be seen as concentrating on different aspects of mythical thought. Structuralism would show that mythical symbols are not (too) overdetermined and fluid in their meanings, and that subjective factors can be modelled in scientific terms. But it would not eliminate the possibility that the symbols themselves are confused or conflated with what they represent. It is perfectly possible to have something of a logical structure to the connotations of mythical thought while still having the people creating, perpetuating and believing these myths unaware of such a structure, and unaware that the myths are symbols of it. In fact, Lévi-Strauss admits that communities believing in the mythical–magical mode are most likely unaware of the structures he imputes to these myths.[13]

It is the lack of awareness of symbols, and their confusion with what they represent, that makes intelligible the peculiar characteristics of magic and myth in general, and of space in magic and myth in particular. Until further work on symbolism more clearly demonstrates the relationship between the fluidity of myth and its structure, we shall rely on Lévy-Bruhl's thesis and the implications that mythical–magical meanings are fluid and overdetermined as the basis of our analysis of space in this mode of thought.

Mythical–magical space

From this point of view, myth and magic seem to emphasise the physiognomic qualities of the world, and everything in mythical consciousness is imparted with life and connected in one universal whole. It is natural that in myth, the model of such a world should come from the most intimate and familiar whole we know, the human body. Hence, the mythical world is seen as an organism with functions and structures analogically similar to a human being's; an organism which often is seen as the mother of man.

As man has breath, blood and a soul, so too must there be something coursing through the world, interconnecting, coordinating and directing its parts. This vital force, or 'cosmic breath', provides the matrix for the mythical world body. In the Melanesian culture it is the *mana*, in the Algonquin it is the *manitou*, in the

Iriquois it is the *orenda* and in the Sioux it is called the *wakanda*.[14] This breath, as it were, is part object, part force. It surges about those objects and relationships which are seen during heightened physiognomic perception to create extreme and profound emotional oppositions such as attraction and repulsion, joy and grief, the known and unknown, permanence and impermanence. This cosmic breath and its manifestations in objects and relationships and the attendant emotions coalesce to produce the primary feeling tone in myth and magic, the feeling which is commonly called the *sacred*. The sacred is mythical–magical power. It emanates from the physiognomically striking elements of the world. In the mythical–magical mode, what is most physiognomically striking, most forceful and potent, is also the most real.

The flow of mythical power courses among objects and relationships according to their physiognomically perceived similarities or similitudes. In what may be termed a law of this mode, like things affect each other: they are sympathetic; unlike things are antipathetic. The similarities and dissimilarities are not based on objectively defined characteristics or relationships alone. They also derive from physiognomically perceived similarities. Therefore, 'mythical thinking in general knows no purely ideal similarities but looks upon any kind of similarity as an indication of an original kinship, an essential identity'.[15] This is most clearly revealed by the peculiar role that space and spatial relations play in the mythical–magical mode. 'The mere possibility of coordinating certain spatial totalities part for part suffices to make them coalesce. From this point on they are only different expressions of one and the same essence, which can assume entirely different dimensions.'[16] This peculiarity, coupled with the magical–mythical view that a thing can be created or affected by the creation or implementation of its symbol, makes 'mythical thinking seem to negate and suspend spatial distance. The distant merges with what is close at hand, since the one can in some way be copied in the other.'[17] This fantastic quality of space has come to be the trademark of the mythical–magical mode. Yet, beneath the mutability of spatial relations and the subjectivity of similitudes, there seems to be an underriding regularity to the relationships between space and causality in magic and myth. This regularity is revealed by the way spatial relations help to engender similitudes.

Spatial relations affect mythical–magical similitudes in two fundamental ways which we will refer to as 'principles'. First, as has been noted, similarities may arise through proximity or contact, constituting what Mauss has called '*the principle of contiguity*'.[18] Two objects in contact or even in close proximity, will share a sympathy which continues when the two are separated so that one object affects the other at a distance. The two factors to bear in mind are these: that things in contact share the potential for cause and effect (that spatial contact or near contact is a sufficient criterion for cause and effect), and that once these things which were in contact are separated, they still affect each other at a distance. This action-at-a-distance principle has become popularised in such expressions as 'Zap! you're dead.'

Original proximity or contact can be thought of as forming a whole. One part of the whole, the part which is removed, becomes a symbol of the whole and because of the confusion of symbol with referent, it becomes the whole (*totem ex parte*). Thus, for instance in a magical context, 'Teeth, saliva, sweat, nails, hair represent a total person, in such a way that through these parts one can act directly on the individual concerned, either to bewitch him or enchant him.'[19] The law need not be confined to biological wholes and parts, nor to living things. 'Everything which comes into close contact with the person – clothes, footprints, the imprint of the body on grass or in bed, the bed, the chair, everyday objects of use, . . . all are likened to different parts of the body.'[20] Moreover, the contact need not be permanent or frequent, and may be imagined. The most important spatial consequences are that contact is a sufficient (but not a necessary) condition for cause, and that cause and effect then can act at-a-distance.

The second principle, *mimetic sympathy*, refers to objects sharing usually visible spatial or geometric similarities and affecting one another at a distance, so that 'a simple object, outside all direct contact and all communications, is able to represent the whole'.[21] Graven images, voodoo dolls, talismans and masks are examples of this principle. In these cases the similitude is based on the sharing of shapes and spatial configurations, though such likeness need not be precise. There need only be schematic representations to produce a physiognomic similarity. Even a 'poorly executed ideogram' will do.[22] As with the principle of contiguity, in mimetic sympathy action also occurs at-a-distance, the mask or talisman taking the place of the thing it represents.

Mimetic sympathy and contiguity in conjunction with the mythical–magical characteristic of conflating symbol with referent, leads to another extremely important characteristic of space in this mode – the attribution of significance to place, independent of substance. A substance which once may have created awe and which since has disappeared, or for that matter may never have existed, as in the occurrence of a miracle, imparts significance to the location, so that the place of the occurrence becomes sacred or holy. Since all objective traces of the event are gone, the explanation for the significance of such a place is that the people are using the place or location as a symbol of the event and are confusing the location, as a symbol, with the event. It is in this respect that myth and magic impart significance to place which goes beyond the scientific meaning that place can have and which, as we shall see in Chapter 7, becomes extremely important in a people's attachment to place, especially at the societal level.

The principles of mimetic sympathy and contiguity are bound together in their establishments of sympathies and antipathies; the first establishes them on the basis of contiguity (whole and part), the second on the basis of similarity (where 'the image is to the objects as the part is to the whole'). Both assume that action can occur at-a-distance. Spatial contiguity, spatial analogy, action at-a-distance, and the significance of place independent of substance are the foundations for myth and magic's most characteristic aspect, its apparent defiance of the normal scientific concept of causality. The scientific concept assumes among other things that causes do not travel through empty space (no action at-a-distance) but occur through intervening substances; that contact and similarity of shapes are not sufficient cases for cause and effect; that similarity of shape is not even a necessary condition; and that place itself has no significance apart from the substances located there. We say that the magical use of space does not coincide with that of *normal* science because as we have seen in the more abstract realms of theoretical physics, and in some areas of the social sciences, there are differences of opinion about the conception of space and its effects on behaviour.[23]

Magic and myth, according to the two principles, establishes similitudes on the basis of spatial relations. From the outside it appears that spatial properties themselves, such as shapes, forms and relative position or location, attain a significance independently of the substances of which they are composed. Space attains a significance which is beyond the relational concept to explain. In a sense, space and its properties appear to be absolute, but the space of myth and magic is not really an absolute conception, because in this mode space and substance are not conceptually separated in a sophisticated sense. Rather, the conception of space in magic and its apparent absolute view is better characterised as *pre-relational*. A spatial property such as a shape is abstracted and separated from substance only long enough to use it as a symbol for that substance or for a feeling; and then, because such distinctions between symbols and referents are never held for long or very far apart, it fuses and confuses them, so that the symbol is the referent. This is how shapes and places come to have significance in myth and magic and how action occurs at-a-distance. It is also how myth and magic conceive of the space–substance axis of the conceptual surface.

The use of spatial relations in this manner to form similitudes creates an enormous chain of influences in which

> both individuals and objects are theoretically linked to a seemingly limitless number of sympathetic associations. The chain is so perfectly linked and the continuity such that, in order to produce a desired effect, it is really unimportant whether magical rites are performed on any one rather than another of the connexions.[24]

But not all in myth and magic is chaotic. Through the specifics of the similitudes established in each magical system, the chain of influences is interrupted, and channelled to affect the things desired.[25] The flow of sympathetic associations is further constrained by the overwhelming geographic regularity and fixity of the most physiognomically arresting objects. In this way the practical perception of a stable space–time system, which originally gave rise to the physiognomic view, reasserts itself by constraining and moulding mythical fluidity.

The flux is like a veneer over the perception of a stable space–time manifold populated by physical objects. Often the most geographically permanent and regular features, such as the sun, moon, stars, mountains, rivers and sea become the most physiognomically arresting. Perhaps the most fundamental and universal projections of feelings on to the world follow the geographical trajectory of the sun. Associations with the sun such as light, warmth, life and birth, are placed in the sun's path, especially in the east and above. Their opposites such as darkness, cold and death, belong to the west and below. In the mythical–magical mode this means that directions actually are thought of as the places or 'homes' of these feelings and functions. The east is the land of light, the west the land of darkness. To the major cardinal directions may be added local physiographical elements such as mountains, rivers, seas and other arresting landscape features upon which mythology is nurtured. This is seen clearly, for example, in the way the sun, the Nile and the desert are interrelated in the Egyptian world view.

The northward flowing Nile provided Egypt with a lifeline amidst the vast and barren desert and became a dominant symbol in the Egyptian world view. The blue river and greenery it supports form a long band of colour through an interminable desert until the river reaches the vast deltas in the north. In the deltaic areas of the Nile, the river diffuses into fans, and no longer provides a visibly prominent landscape feature. Here, the most visually arresting feature is the motion of the sun over the vast undifferentiated flatness of the delta. In ancient times, this area was part of the Lower Kingdom and its religion was dominated by the sun, while the religion of the Upper Kingdom, in which there was a clear river channel, was dominated by the Nile. Later, the solar-dominated theology of the north was added to the Nile-dominated theology of the south. The merging of the two theologies established a strong sense of symmetry and order which was reflected in Egyptian architecture, literature and language.

East and west, the regions of the desert, were thought of as barren and inhospitable. Yet the east was better than the west, because it was the place of the rising sun. The major orientation of the Egyptians was south, to the source of the Nile. The Egyptian

took his orientation from the Nile River, the source of his life. He faced to the south, from which the stream came. One of the terms for 'south' is also a term for 'face'; the usual word for 'north' is probably related to a word which means the 'back of the head.' On his left was the east and on his right the west. The word for 'east' and 'left' is the same, and the word for 'west' and 'right' is the same.[26]

The metaphorical use of spatial expressions, we noticed, helps language 'place' abstract concepts such as time in a system of relationships, making them more accessible to human consciousness. In the same way the fixed geographic features, anchoring the mythical world to reality, provide a spatial matrix for new mythical conceptions which are placed, and thereby reified, in areas of the world which seem to possess feeling tones sympathetic to those of the mythical concepts. The Keresan Pueblo Indians, for instance, believed in a symmetrical cosmic landscape. The earth was a square and in each corner was a house in which lived a god.

In the northwest corner was the House of Leaves, the home of Tsityostinako, or 'Thought Woman'; she could cause things to happen merely by thinking of them. Spider Grandmother lived in the House of Boards in the southwest corner. Turquoise House was in the southeast corner; Butterfly lived there. And in the northeast corner lived Mocking Bird Youth in Yatkana House. . . . [In addition to the four cardinal points the zenith and nadir were also important.]

Each of the six directions had a color; the north was yellow, west blue, south red, east white, zenith brown, and nadir black. At each of the cardinal points lived a god. Shakak lived at Kawestima, the north mountain; he was the god of winter and of snow. Shruwitira, a man-like god, lived at Tspina, west mountain. A gopher-like god named Maiyochina lived at Daotyuma, or south mountain; he helped crops to grow. Shruwisigyama, a bird-like god, lived at east mountain, a fox-like god at the zenith, and a mole-like god at the nadir.[27]

In addition to these, the Keresan Pueblo Indians had other gods placed in homes on their landscape.

Geographical locations themselves assume a far greater significance in the mythical–magical mode than can be explained by the relational concept. In myth and magic,

the distinction between *position* and *content*, underlying the construction of 'pure' geometric space, has not . . . been made and cannot be made. . . . Hence . . . no 'here' and 'there' is a mere here and there, a mere term in a universal relation which can recur identically with the most diverse contents; every point, every element possesses, rather, a kind of tonality of its own. . . . [E]very position and direction in mythical space is endowed as it were with a particular *accent*. . . .[28]

Space is not conceived of as isotropic in this mode. Its parts or subdivisions would not follow the functional relationships of geometry or mathematics which can simply, by extending the parameters, extend the space indefinitely. Rather, the space of myth and magic is non-isotropic because it is structured by rules of inherence.[29] Through *pars pro toto* each part potentially contains the whole and hence is the whole.

Objects of awe are seen to be indissolubly linked to the place in which they occurred. Because the place and the objects are not conceptually separated, the place can symbolise the object, or, in the mythical–magical view, be the object. An object of awe can be removed from a place or even disappear as in the case of a miracle which may have occurred at a place only once and then only briefly. The power of this occurrence, though, is enough to have its effect reside in the place and make it holy. The rule of *pars pro toto* allows each holy place to be a microcosm of the whole, to be a recreation in miniature of the world. This fusion between space and substance in conjunction with the unsophisticated symbolisations, infuse geographic locations, geographical directions and other spatial properties with enormous significance.

Because shapes are significant in this mode, the construction and placement of potent designs can make any site sacred or powerful, and the power of a holy place can be enhanced by making it more clearly demarcated from other places and by building cosmic symbols on the consecrated ground to enshrine it. Structures with magical shapes, as symbols of the cosmos, draw cosmic powers down by creating a microcosm.

Making a settlement in many technologically simpler societies is equivalent to laying out the geographic form and extent of a society. The physical act of building a settlement becomes a focus of mythical–magical concern with place. The act of settlement is a re-enactment of the mythical creation of the world.[30] It is in the form of the built environment and its relationship with the rest of the inhabited area, that we see the most striking evidence of mythical–magical beliefs affecting the geographic landscape.

We noted that the Hottentots constructed their huts and their settlements to recreate the cosmic form. Other tribal societies, such as the Pawnee Indians of the American Plains, did the same; the patterns of the tepees represented the stars in the heavens.[31] Attempts to recreate the cosmos on the ground were made in more advanced societies as well. In India, cities were supposed to be designed according to a mandala replicating a cosmic image of the laws governing the universe. Manuals of architecture defined the shape of this mandala as a square which can be divided thirty-two ways into smaller squares or quarters.

> All existence is reflected in this magic square. It is an image of the earth, which is a square derived from a circle; at the same time it is also the sacrificed body of the primeval being, Purusha. Man and earth correspond to one another in this image. Time enters the mandala by co-ordinating the signs of the zodiac, and space does so by orienting the square towards the four . . . cardinal points.[32]

Each square subdivision or *pada* was the location of an individual god. The centre square was occupied by Brahma, and each ring of squares was occupied by lesser gods as the distance from the centre increased. This was the principle behind the design of Indian cities. Because Indian cities were constructed from extremely perishable materials such as unbaked clay and wood, it is difficult to determine how closely these prescriptions were actually followed.

Propitious sites for the location of settlements in China were determined by studying 'earth currents' which revealed themselves through local relief. Such divination was called *geomancy* (*Feng-Shui*) or, literally, 'winds and waters', the two elements thought to be the principle agents shaping the landscape. Geomancy is 'the art of adapting the residences of the living and the dead so as to cooperate and harmonize with the local currents of the cosmic breath'.[33] No Chinese City was planned without the advice of a geomancer.[34] Like all magical systems, the particulars of geomancy vary depending on the conceptions of different adepts. In general, geomancy assumed that the earth is a live, breathing organism. 'When it moves its breath produces the *yang* or male energy, and when it rests its breath produces the *yin* or female energy. Nature's breath is a two-fold element consisting of *yang* and *yin* energies which interact continuously and produce all forms of existence on earth.'[35] The cosmic breath follows earthly channels in a way comparable to the flow of blood in the human body. These channels are revealed by local relief. For instance, *yang* energy runs through the lofty mountain ranges while *yin* energy runs through the lower ranges. The object of geomancy is to find the combination of earth currents at a place which would be propitious for specific kinds of human activities. Geomancy was used to locate Chinese cities and Chinese cities were conceived of as cosmological symbols. The heavens were represented by a circle and the earth by a square. According to Chinese ritual books, cities were to be square in shape, oriented to the cardinal directions and enclosed by walls with twelve gates, one for each month. Three gates were to be placed in each wall, opening the city to the cardinal directions. Within the city there was to be a walled square for the residence of the monarch. North of this square was the public market. To the south of the square on the east and west sides of the principal street which ran from the southern gates to the palace were 'two sacred places: the royal ancestral temple and the altar of earth'.[36] The centre of the city and focus of the cosmic symbol were the royal halls in which the ruler, when conducting affairs of state, faced to south, the favoured direction in Chinese cosmology, (see Figure 6.1).

As the centre of the kingdom, the city was essential to the countryside. The heavenly effluvia attracted to the city was to flow through the twelve gates, spreading outward through the country, fructifying the land. Indeed, the gates of Chinese cities were often larger than they needed to be for normal commerce, because they were to provide ample room to allow the cosmic forces to flow through them.[37]

Mythical—magical views of settlement and landscape were not limited to the non-Western world. Traces of it are found in the West as well. We can suppose that the tribal communities which inhabited Europe before civilisation had their share of magical conceptions of nature. But all that we know of their cosmologies, settlement patterns and land use have to be inferred from scanty evidence of the records of Greek and Roman travellers and a few archaeological remains such as the megaliths of north-western Europe. Apart from these there were no large architectural undertakings and there was little or no evidence of urban settlements before the influence of Greece and Rome.[38] In the long history since Roman colonisation of Europe, almost all traces of original tribal culture on that continent have been obliterated.

But the Greek and Roman conceptions of settlement and landscape did include mythical—magical elements of their own. Each Greek city, for example, had its gods or deities which were symbolised in the city altar or *prytaneum* within the *acropolis* or sacred core of the

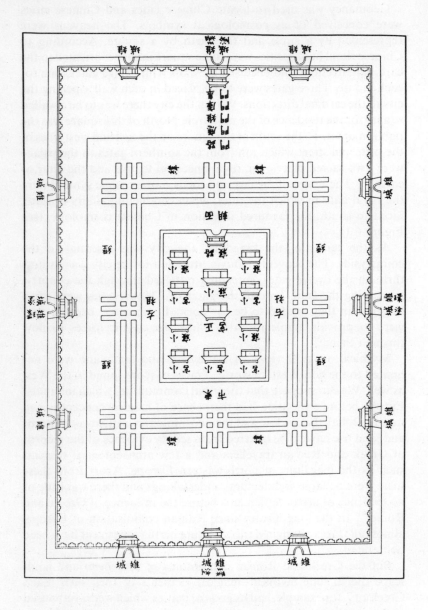

FIGURE 6.1   *The stylised version of Giwang-d̂ˌiĕng (Wang-Ch'eng)*

city.[39] Rome, too, had its gods and the city itself was supposed to have been laid out by Romulus according to cosmic plan.[40] Such demarcations were later to be incorporated in Roman town plannings. But unlike any other civilisations at the time, the Greeks and the Romans possessed the beginnings of scientific attitude which became the essential criterion in differentiating Western from non-Western societies. This attitude altered and suffused many of the mythical-magical components of the Graeco–Roman world view.

Christianity introduced its own conception of the world. But in addition to the conventional Christian world view, there developed a magical or occult tradition about which the church was extremely uneasy. This tradition was most clearly expressed in the late Middle Ages and Renaissance when there was a resurgence of interest in classical thought and also in science. The close association of Western magic with religion and science makes it one of the most interesting magical systems for our purposes.

The world view of the late Middle Ages and Renaissance corresponds in many ways to the mythical–magical mode. This age viewed the objects of nature as reflections of sympathies, antipathies and chains of influences, based on the resemblances of things to one another. 'The universe was folded in upon itself; the earth echoing the sky, faces seeing themselves reflected in the stars and plants holding within their stems the secrets that were of use to man.'[41] Things formed a teeming abundance of resemblances, reflecting, linking and echoing one another throughout the universe. This teeming pattern would have no beginning or end, or direction, without a visible key to interpret the flows and connections. This key was provided by the *signatures* which stand out as visible marks or signs left by the Creator's hand to help man unravel the resemblances of the universe.[42]

Simply as a means of understanding God's design, the signatures were a special form of the ancient theophany which maintained that God's presence and purpose is revealed in his creations.[43] The signatures on earth, reflecting the will above, form letters in a book of nature to supplement the revealed words of God. Often the meaning of the signatures rested on easily visible similitudes. For example, a stamp of a special astral image might be found on an object, 'so that if you cut across the bone of a solar animal or the root or stem of a solar plant, you will see the character of the sun stamped upon it'.[44] This viewpoint, found in Christian thought as early as the fifth century, became part of a rich tradition of nature imagery in Christian theology.[45]

If one knew precisely what was being reflected and how it could be read, these signatures could be copied in man's artifacts to magnify and channel the heavenly effluvia in order to control the environment. To help determine the signatures, more explicit clues were needed than were found in the Bible, so attention was turned to the remains of ancient pagan religions, in occult documents which were thought to have ancient and even pre-Biblical sources, and to the principles of astrology and alchemy. Through this search the relationship of space to action became interlaced with the complex threads of Neo-Pythagoreanism and Cabalistic number magic.

From the Pythagoreans came the transformation and expansion through analogy of the visible geometric shapes to the realms of numbers and sound. For example, the number one was thought to be a point, having position and magnitude. It was identified with wholeness, reason and the limited. Two was a line, and suggested diversity and the unlimited. Three was a surface and four a solid, which signified the composition of matter. Point, line, surface and solid were the four possible dimensions of all forms, and there were also four elements of which all matter was composed.[46] In addition to the association of spatial properties with numbers and with physical and philosophical virtues, the Pythagoreans added the specific link between numbers, shapes and sounds (and also reinforced the primacy of four) through the discovery that musical intervals depend on arithmetical ratios of lengths of strings at the same tensions; 2:1 gives the octave, 3:2 the fifth and 4:3 the fourth, all of which are based on four numbers.

An equally important part of the Pythagorean legacy was the suggestion that a universal harmony is recognisable and translatable simultaneously into numbers, sounds, shapes and matter, so that all the world is number. In the Western magical tradition of the Renaissance, this suggestion existed side by side with the complementary scheme of the *Cabala*, although it is not clear if any hermetic took complete advantage of the complementarity to form a single system of magical principles Nevertheless, the potential was there. The Cabala, among other things, supposed the world to be constructed upon numbers and letters.[47] In Hebrew, letters also stood for numbers, and the Cabala, through the rules of the *Gematria*, elaborated the connection between numbers, letters, words and sentences. Through this technique the numerical value of words and sentences could be determined, and thus a further dimension was added to the hermeneutics of God's words.[48] The Cabala and Neopythagoreanism together provided the potential for the interconnection and interchangeability of numbers, shapes, sounds, letters and words into a powerful magical language. Statements made in one mode of expression would simultaneously be made in the other, and underlying all would be a celestial harmony.

The application of magic in the West to enhance, control and alter the environment was far more modest than its theoretical formulations would have portended, in part because magic, though it fitted within the Renaissance world view, was never completely accepted by the church.

The unambiguous use of shape for magical purposes is clearest on the smallest scales, as in signatures on magical charms. A category of sympathetic magic, called the *Doctrine of Signatures*, used signatures from nature for medicinal purposes.[49] This doctrine asserts that the resemblance of a plant to parts of the body or to symptoms of diseases signifies its appropriateness as a treatment for those parts or ailments.[50] Hollow-stalked garlic was used for the windpipe, the throatwort was for bronchitis, and liver-shaped liverwort for the liver. 'Lungwort, spotted with tubercular-like scars, was used for consumption . . . (and) the Lily of the Valley whose flowers hang on its stalk like drops – for apoplexy, a disease caused, they learned, "by the dropping of the humours into the Ventricles of the brain." '[51]

On a larger scale, we have evidence that rooms, houses, memory theatres, theatres and churches were in part deliberately designed as talismans to be in sympathy with the heavens and to magnify and channel the effluvia. The effects of viewing magical shapes on the observer's memory and imagination was considered an important element in directing the heavenly forces. Visualisation played' a prominent role in the contemporary theories of memory and cognition, as well as in magic, where channelling the effluvia may have required the operator to visualise (in his mind's eye) the sympathetic signatures. Architectural structures were designed on magical principles to aid memory and imagination, and to enable men to alter their environments and to create beautiful art, poetry and music in perfect harmony and proportion. Magical shapes permeated the entire form of certain structures as well as their embellishments. Renaissance mystics followed the Roman architect

Vitruvius, on his design of the theatre, in which much attention was given to the role of astrology. They designed theatres and memory rooms (or memory theatres) in order to have memory (and thought) respond to the forces of the universe.[52]

It is rare to find in the West, cities designed according to mythical–magical principle. A few cities incorporated well-known hermetic shapes and patterns in their ground plans, but it is difficult to determine whether such shapes are simply culturally accepted and aesthetically pleasing styles, or if they were designed intentionally to act as giant signatures to tap astral forces. The medieval French town of Brive, for instance, has, after the seven heavenly bodies, 'seven gates, seven concentric roads, seven streets radiating from the cathedral center, and seven parochial quarters'.[53] We must turn from the ground to the 'idea' of theoretical and Utopian towns to be certain of magical intent at the scale of cities. For example, Campanella's City of the Sun was an intricate model of an ideal city whose structure was to serve as a talisman, capturing the heavenly effluvia and channelling them to the inhabitants through a range of magical principles.[54]

The City of the Sun was supposed to be circular, built upon a hill and divided into seven rings named from the seven planets. These rings were intersected by four streets which passed through four gates oriented in the cardinal directions. At the top of the hill, within the seventh ring, was a circular temple containing a globe of the earth and of the heavens. On all the walls of the seven rings were decorated images representing the essential knowledge of the universe. The temple and the city's walls with their images and emblems were to capture the astral forces and serve as a giant memory device and a book of nature. The boys of the city would learn their mathematics, alphabets and languages by seeing these represented on the walls, and the women selected to become mothers would often gaze at the statues and pictures of heroes.[55]

The largest coherent signatures of man's creation during the Middle Ages and Renaissance were coincident with the walls of the city. Far less attention was given to the countryside as a possibility for human design. Although a book of nature and influenced by astral forces, the spatial configuration of the countryside, for the most part, seemed beyond man's control.

According to modern mystics though, there may have been a long-forgotten age when magic, as the supreme power and the only world view, was able to ring the earth with chains of talismans to control the astral and telluric forces and use and reshape the configurations of the countryside. Renaissance magic would be only an incomplete resurgence of this ancient occult tradition which once supported a complete working system of immense power at all scales. Proponents of this position argue that the powers were lost because of some holocaust, and so too were most of the visible structures. They point to megalithic monuments and wonders of the ancient world, like Stonehenge and the pyramids, as the last remains of the once universal system of talismans.

Only part of the key to the ancient mysteries would be the visible signatures on the earth's surface. As the Neopythagoreans and the Cabalists remember, the complete answer would be the lost perfect language which simultaneously expressed the universal harmony in letters, numbers, sounds and shapes of a complete environment that could be known in thought and created through action. So harmonious would be this key to nature that thought would flow to action, and the mind in possession of the key would exercise mental power over material objects to move or change them in a far more complete way than was expressed in the sixteenth-century concept of image and memory.

These thoughts about a grand magic controlling the environment in a lost age stand naked and vulnerable in the light of twentieth-century science and technology. So, too, do the few disjointed remains of the hermetic traditions, such as twentieth-century astrology and chiromancy, whose practitioners are more often than not unaware of how eviscerated their art has become. The meanings of such systems of magical symbols seem conventional and arbitrary. This is precisely what symbols in this mode are not supposed to be in societies in which myth and magic are integral and vital elements. But symbols which may seem transparent, natural and self-explanatory from within primitive societies where the mythical–magical mode does not compete with sophisticated and fragmented patterns of thought such as science and art, are, because of their physiognomic basis, susceptible to multiple and contradictory interpretations by those who are not part of the culture and who have not been indoctrinated into this system of thought. Thus far, the way to see transparency of symbols is to see through the eyes of the culture, to see the mythical–magical systems for the society not as symbols but as reality. This may only be possible to one who is a native of that society and who believes in the society's myths and magic. If we look at the symbols analytically, as we must from a scientific perspective, or even from the sceptical perspective of the Westerner, we then can only sense that these symbols are transparent. Determining exactly how and why they are transparent would require that we identify natural and universal symbols and that we unpack, scientifically, the fluidity and overdetermination of meanings – a task which thus far has proven to be intractable.

Part IV
Unsophisticated and Sophisticated Patterns

7
The Societal Conception of Space

We have concentrated on the essential symbolic structures of modes of thought that are potentially a part of the intellectual capacities of individuals in all societies. These modes are social in the sense that individuals producing them are influenced by society and the organisations or communities to which they belong. These social conditions undoubtedly affect the contents of the modes and the degree to which they are separate and distinct. There are, however, realms of activities that are predominantly group, collective or social, and in which the individuals act in the name of the group. Nations, cities, armies, families and scientific, religious and artistic organisations are examples of collective or social relationships or facts. These facts are linked to space in two related ways. First, the social organisations and the individuals within them are 'in' space and their interactions have spatial manifestations. Thus we have families in cities, and cities in regions containing other cities, and so on. Analysing these relationships has been the traditional concern of social geography and has resulted in such theories as central place, land use, and the gravity and potential models.

Second, social organisations are often territorial, a fact largely overlooked by all but some political geographers. Territoriality here does not mean the location and extension in space of a social organisation or of its members. Rather it means the assertion by an organisation, or an individual in the name of the organisation, that an area of geographic space is under its influence or control. Whereas all members of social organisations occupy space, not all social organisations make such territorial assertions. The social enforcement (and institutionalisation) of such assertions—in the form of property rights, political territories or territories of corporations and institutions–provide the context necessary for social facts to exhibit the first type of spatial properties. The forms such territorial structures take and the functions they provide depend on the nature of particular political economies.

Both the spatial relations and territoriality of social facts involve distinct conceptions of space which we will refer to as *societal conceptions*. We will explore the societal conceptions of space associated with the political–economic structures of society as a whole.

Mixtures of modes

The greatest differences among conceptions of space at the level of societies as a whole are found when we classify social systems in a general evolutionary schema from primitive societies to civilisations, with some civilisations evolving into modern nation states. (We will not consider the various transitional stages between the primitive and civilised such as chiefdoms.)[1]

There are two primary properties to the societal conception of space which apply especially to the level of the political–economic structures and which most clearly illustrate the differences in views of space that are associated with the differences between primitive and civilised. The first property is the conception that a people have regarding the relationship between their society (or social substance) and its geographic place. As with other things, societies occupy space. The first property refers to a people's conception of this relationship. Societies tend to forge strong ties to the places they occupy and to justify these ties through social organisations and procedures. Different societies conceive of these ties to place differently. In some primitive societies the social order is not thought of as possessing a continuous extension in physical space. Rather, the society is anchored to the earth's surface in very special locations such as holy places, sources of water and traditional camp sites. The intervening areas, although known to the members, may be unimportant to them in a territorial sense. In such cases the territorial boundaries would tend to be vague. For other societies, the social order may be conceived of as extensive over space where the boundaries may be more or less clearly defined and may become territorial. In civilised societies, parts of the society are seen as possessing continuous extension, but what parts and how clearly their boundaries are defined differ from one type to another.

The second property of the societal concept of space is the knowledge and attitude that a people have regarding other peoples and places. In such cases we are more interested in the elaborateness of the spatial viewpoint than in the specific details or content of the knowledge. We can think of the elaboration of such knowledge in a way analogous to the development of the child's conception of spatial relations.[2] We recall that Piaget characterises the levels or stages in the child's development of an awareness of spatial relations as first in terms of a simple topological conception where the child is aware of little more than his own presence, then in terms of a more complex topological schema, and then in terms of projective and Euclidean relationships. This sequence can be used to describe in general terms the levels of social awareness of space. There are, for example, primitive societies that have virtually no knowledge of other places or peoples except their own. Their view is extremely ethnocentric and space is literally the place or territory they occupy. Beyond the place, the idea of space does not apply. Other primitive societies may have a slightly more elaborate view of other people and places. They may be conscious of where other people border them, but, except for this, they may be unaware of their neighbours' territories and of what extends beyond. In civilisations, with the possible exception of feudal societies, we generally find a more articulated view of the territories beyond one's own, a view that conforms to a projective or Euclidean space.

The relationship of these two properties of the societal view of space to our conceptual surface is as follows. With regard to the first property, a society may rely entirely on unsophisticated modes of thought such as mythical–magical views to connect the society to place, or a society may incorporate both unsophisticated and sophisticated modes. With regard to the second property, a society may not be able to attain a sufficient conceptual distance between the world and itself to have more than a conception of its own extension. Such a view would be unsophisticated. On the other hand, a society may be able to conceptually separate itself from the world sufficiently to imagine the existence of others in the world and of the spatial relations that pertain among them and thus attain a sophisticated view of the world.

The two properties are roughly interrelated with regard to their degree of sophistication and these degrees are roughly related to the division of society into primitive and civilised. In societies which do rely heavily on unsophisticated–fused views, the social order is often mythologised and linked to place through a mythical–magical mode. Similarly, the conception of the space surrounding the society is seen through this mode and in fact becomes submerged within it. In civilisations (and all civilisations have elaborate sophisticated modes of thought), both the anchoring of society to place and the relationship of the society to other places can be viewed in a sophisticated pattern. But these societies incorporate unsophisticated modes as well. Which pattern predominates depends on the particular structure of the society and on its relationship to other societies.

The primitive

The term 'primitive' has come to have a bad connotation. Its use is thought to indicate a condescension and arrogance on the part of the presumably non-primitive person using it. Furthermore, it implies at least a two-stage (primitive to modern) process of cultural evolution. The idea of general stages in the evolution of cultures has until recently been spurned by many American cultural anthropologists who professed views of cultural relativism and uniqueness. It is incontestable that the idea of the primitive implies evolution from a single source or type of society in the sense that it refers to a phylogenetically earlier state where the phylum is human society. Whether in fact there was a social system that can be called primitive is another matter and, on this score, the evidence does suggest there to have been a primitive 'type' of society which was qualitatively and quantitatively different from civilisation. But in order to discuss this type clearly we must first point out that the original meaning of the term 'primitive' is in no sense pejorative.[3] It meant primary or original in time and even in rank.

Our use of the term primitive is intended to convey this original non-pejorative meaning. We are interested in the earliest, original or primitive societal views of space.

But what primitive society was like can only be known incompletely from fragmentary and often controversial evidence. Such reconstructions run the risk of being overly simplistic and idealised. We should be especially wary, for the incomplete evidence paints a picture of a society which in many respects is more organically interrelated and attractive than what we know of our own or other civilisations. This picture of a simpler, more attractive society, may be an unavoidable outcome of positing an evolution which, for living things, always turns out to be a progression from the simpler to the more complex.

The kind of society which is referred to as primitive belongs before the rise of ancient civilisations some 7–8000 years ago. There are, of course, no such societies left to observe. Our informed views about them come from archaeological reconstructions and from anthropological field work in preliterate societies which have comparable technologies to those unearthed by archaeology. Even though the 'primitive' societies studied by modern anthropologists have changed in the last 8000 years and have been in contact with more developed societies, their technologies and social structures are radically different from modern ones and closer to what we know of the older societies from archaeological remains. Therefore, with proper precautions, we can use contemporary ethnographic data as evidence for a plausible characterisation of the earlier societies. We will concentrate especially on those features which set the primitive apart from civilisations and which most clearly illustrate their societal view of space.

Primitive groups are less complex than civilisations. They have less division of labour, internal specialisation, fewer numbers and smaller territories. But among them there are different orders of complexity, ranging from the bands and clans of the hunters and gatherers through the more complex tribal societies to perhaps the highest forms of primitive social organisation, the tribal confederation, such as the Iroquois of north-eastern United States or the Maori of New Zealand. In all primitive groups, the family is the basic unit and often the next higher social unit is the band.[4]

For the most part, the hunting and gathering bands are nonsedentary. A band may have a different ecological habitat for each season. Their numbers never approach the size of even a modest town. Of the extant hunters and gatherers, perhaps the Eskimos have the largest villages, numbering in places of good hunting, several hundred inhabitants.[5] But most villages are much smaller. Their primitive technology makes for little specialisation and division of labour beyond that of age and sex. Their nomadic existence makes the family unit the essential core of those societies. Links beyond the nuclear family are established through conceptions of kinships.

The tribe is more complex than the band. The term covers a range of societies occupying different habitats, and having different economics and population sizes. In the tribe, as in the band, the family is the core unit. However, the kinship links in tribes are much more precise and extensive than in bands. Tribal settlements may contain only a single primary family, but the average size of the villages of non-intensive agricultural tribal groups is approximately 1—200.[6] In intensively cultivated areas, tribal settlements could be as high as 1500.[7]

Tribes have a more segmented social order than do clans and bands, but not until we come to civilisations do there exist true economic *classes*. In tribes, the division of labour is still predominantly based on age and sex. Community life in the tribe is family oriented. The community seems to provide its members with an intimate and enveloping sense of belonging. The 'naturalness' of this unit is one of the most striking aspects of primitive societies. The sense of a unified community, as Diamond put it, 'spring[s] from common origins, [is composed] of reciprocating persons and grow[s] from within'.[8] The sense of community is enhanced by the tendency in primitive societies to use the family as an analogy for society and its relationships with the world. This analogical extension of the family contributes to the general primitive view of the unity of nature. It creates a personalism which extends to nature and which underlies, and is perhaps the most distinctive element of, primitive thought and behaviour.[9]

The intimate link between person and community does not stifle individuality and personal expression. In fact, according to many observers, there is greater allowance for individual expressions in primitive societies than in civilised ones.[10]

How the individual sees himself in relationship to the community is difficult to determine. Most observers agree that the bond between individual and society among the primitives is unusual and unusually close. This impression has led observers to believe that in some respects the individuals do not conceptually separate themselves from the group; that is, the conception of individual and group is pre-logical. As Lévy-Bruhl has said of primitive societies, 'the individual as such, scarcely enters into the representations of primitives. For them he only really exists insofar as he participates in his group.'[11]

As we have seen, it is wrong to argue that the primitive's mind is inherently different from anyone else's (and this was not what Lévy-

Bruhl was saying). Rather the issue is about the actual expression or form the mind employs – in this case when symbolising the relationship between individual and society.[12] Radin, for one, does not think that the unusual relationship of individual to group is due to a lack of conceptual separation between the two categories. Primitive man, according to him, is perfectly able to separate the two in the context of social thought and in other realms as well. Yet, the explanation he offers for the closeness of the individuals to the community is not far from Lévy-Bruhl's position. In Radin's explanation, we find an appeal to a larger entity, to which the parts belong, as a means of unifying the parts and yet keeping them conceptually distinct.[13] But seeing two things as distinct on some occasions and fused on other occasions is very close to what Lévy-Bruhl was saying. In addition, Radin has little to add about the nature of the configuration and the conception of society within it. 'What the real and essential nature of this social configuration consists of . . . is well-nigh impossible to determine.'[14]

Although Radin's position does not offer a clear alternative to Lévy-Bruhl's thesis about why individuals and society appear to be so closely associated in the primitive world, his view that they hold the individual and the group to be distinct concepts would be strengthened by implication if he could substantiate his more general contention that in many realms of life primitives approach problems abstractly. If they do, there would be no compelling reason why they would not conceptually distinguish between individual and society.

As the title of his book, *Primitive Man as Philosopher*, suggests, Radin believes that the primitive mind often indulges in the highest abstractions, that the primitive can think philosophically. If we take the definition of a philosopher here to be someone 'who struggles with intellectual problems in abstract terms and offers solutions of some originality', it seems that Radin does not have convincing proof that primitive societies possess such individuals.[15] The evidence which Radin and others have assembled does indicate that primitives are able to think as abstractly as anyone. About that there is no doubt, but it also indicates that by and large they do not find such thoughts either desirable or useful. In fact, they try to avoid them.

It is their wariness about and avoidance of abstractions that more than anything else are the essential characteristics of non-mythical–magical primitive thought. Many of their so-called philosophical statements are in fact admonitions against abstractions. We in the

West, for instance, are expected to love, and to believe that love is a good thing. The primitives tell us not to love love, but to love specific people, and to demonstrate such love rather than pronounce it.[16]

The primitive's tendency to shy away from abstractions may pertain especially to his thoughts about society because the primitive social order does not present the need for social abstractions. Primitive society lacks such powerful elements as economic inequality and class conflict which would create social differences that only abstract social philosophy or revolution would help to reconcile. There are no legal, political, administrative institutions, organisations or apparatus of state apart from and above the people. Primitive society is participatory. Conflicts in society are not directed against institutions or corporate entities, but against specific individuals. Their arguments are not abstracted into political theories which offer alternate conceptions of social order. When an individual finds a social rule or norm to be an insurmountable obstacle, he may break the rule, or leave the group to form a new group wherein the rules of conduct may very well be a copy of ones of the society from which he left. The rule is not seen as an obstacle for the attainment of a particular goal and may never be an obstacle again. Because conflicts are personal and not abstract, such societies do not produce political and social alienation. Opponents are people, not institutions or group entities. Opposition is not between 'we the people' and 'they the society'. There simply are no revolutions in primitive society.[17] In Diamond's terms,

> The primitive stands at the center of a synthetic holistic universe of concrete activities, disinterested in the causal nexus between them, for only consistent crises stimulate interest in the causal analysis of society. It is the pathological disharmony of social parts that compels us minutely to isolate one from another, and inquire into their reciprocal effects.[18]

The lack of need for abstractions about society at the primitive level explains the scanty and problematic evidence of its occurrence. It lends weight to the view that in primitive society the individual and society are not thought of as very distinct elements, because there is no need for such abstractions in primitive societies.

The organic relationship between individual and society is recapitulated in the relationship between society and milieu. As the individual is not alienated from society, society is not conceived of as independent of the place which it occupies nor are individuals alienated from the land. A constant and intimate knowledge of place enveloped by a mythical view of the land fuses the society to place. Place is often inhabited by the spirits of the ancestors and a specific place may have been given to a people by their gods. In Australia, each totemic group is associated with a place from which the totemic ancestor is supposed to have emerged. When a person dies his spirit returns to the place of his totemic origin. The Northern Aranda aborigine of Australia

clings to his native soil with every fibre of his being. He will always speak of his own 'birthplace' with love and reverence. Today, tears will come into his eyes when he mentions an ancestral home site which has been, sometimes unwittingly desecrated by the white usurpers of his group territory . . . mountains and creeks and springs and water-holes are, to him, not merely interesting or beautiful scenic features . . .; they are the handiwork of ancestors from whom he himself has descended. He sees recorded on the surrounding landscape the ancient story of the lives and the deeds of the immortal beings whom he reveres; beings who for a brief space may take on human shape once more, beings, many of whom he has known in his own experience as his fathers and grandfathers and brothers, and as his mothers and sisters. The whole country-side is his living, age-old family tree. The story of his own totemic ancestor is to the native the account of his own doing at the beginning of time, at the dim dawn of life, when the world as he knows it now was being shaped and molded by all-powerful hands.[19]

Physiognomically arresting landscape forms are often the ones incorporated into the myths, helping to anchor the society to place. According to Penobscot Indian lore, much of the landscape is a result of the peregrinations of the mythical personage Gluskabe. The Penobscot river came to be when he killed a monster frog, Gluskabe's snowshoe tracks are still impressed on the rocks near Mila, Maine. A twenty-five foot long rock near Castine is his overturned canoe, the rocks leading from it are his footprints, and Kineo mountain is his overturned cooking pot.[20] A place on the earth in many creation myths was given to a people specifically by the gods. The Pawnee, for instance, believe that they were guided from within the earth to their present place by Mother Corn, and the Keresan Pueblo Indians believe that they were led by Iyatiku, their mother, from the centre of the earth to a place on the earth's surface called Shipap.[21]

Belief in the inhabitation of the land by the spirits of ancestors and in the mythical bestowal of the land to the people have occasioned a powerful communal sense of ownership and use. To have access to the land one must be a member of the society, which means partaking in the spiritual history of the group. For example, in Bakongo tradition,

> the ownership of the soil is collective, but this concept is very complex. It is the clan or family which owns the soil but the clan or family is not composed only of the living, but also, and primarily, of the dead; that is the *Bakulu*. The *Bakulu* are not all the dead of the clan; they are only its righteous ancestors, those who are leading a successful life in their villages under the earth. The members of the clan who do not uphold the laws of the clan . . . are excluded from their society. It is the *Bakulu* who have acquired the clan's domain with its forests and rivers, its ponds and its springs; it is they who have been buried in this land. They continue to rule the land. They often return to their springs and rivers and ponds. The wild beasts of the bush and the forest are their goats, the birds are their poultry. It is they who 'give' the edible caterpillars of the trees, the fish of the rivers, the wine of the palm trees, the crops of the field. The members of the clan who are living on the soil can cultivate, harvest, hunt, fish; they make use of the ancestral domain, but it is the dead who remain its guardians. The clan and the soil it occupies constitute an indivisible thing, and the whole is under the rule of the *Bakulu*. It follows that the total alienation of the land or a part of it is something contrary to Bakongo mentality.[22]

Society and place were so closely interrelated that for the primitive to indulge in speculation about the society elsewhere or about the society having a different spatial configuration, would be like severing the roots from a plant. It could be of no value. But such intellectual contrivances are precisely what social planning and theory require. Statements like 'what if the social order were altered so that land were held differently'; 'what if the village were redesigned, placing this here rather than there, making that rectangular rather than circular, so that certain goals will be more easily attained', are the basis of a conceptual approach to society and place. They involve the conceptual separation and recombination of social activities or substances from space – a separation which underlies all forms of social theory.[23] This separation and attempted recombination of space and society are absent in the primitive world. The place and the people are conceptually fused. The society derives meaning from place, the place is defined in terms of social relationships, and the individuals in the society are not alienated from the land.

Civilisation

There has been no comprehensive and empirically verified theory of how (some) primitive societies gave rise to civilisations. We know that in general there must have been an agricultural surplus and a system of distribution which would allow the surplus to support non-agricultural specialists in an urban environment. But how such surplus and distribution came about, the forms they took, and the reasons for them remain unclear and most likely variable from place to place. In some cases, social forms of greater complexity than tribes might have come about from unusual environmental pressure such as flooding or desiccation. In others, it may have been the ascendency of a priestly class, or propitious locations for trade that provided the impetus. What makes the rise of civilisations especially problematic is that from our reconstruction, primitive society seemed not to have had internal contradictions: it was in equilibrium. Yet from some of these societies came the complexities and contradictions of civilisations, and different kinds of civilisations at that.[24]

While the reasons for that transition from primitive to civilisation are unclear, it is well established that in all of the changes there was a replacement of the classless society by a class society. Civilisations have economic classes. The factor of economic class is extremely important for it leads to different and often contradictory and antagonistic views of social order.

No longer was society seen by all of its members in the same light. Rather, it became a multifaceted and ambiguous concept. In order for civilisations to cohere, such different views would have to be reconciled. The mechanism concurrent with the formation of civilisation which served to reconcile these problems was the formation of the state, whether it be the archaic state, the feudal state, the oriental state or the modern state.[25]

The state stands as an institution above the people encompassing the entire society. Its parts seem not to be equivalent to the citizenry. Unlike primitive society, the government of the state and its officials and power are distinct from the people and their powers. The government has the power to coerce its citizens. The state, as Engels explains,

> is a product of society at a particular stage of development; it is the admission that this society has involved itself in insoluble self contradiction and is cleft into irreconcilable antagonisms which it is powerless to exorcise. But in order that these antagonisms, classes with conflicting interests, shall not consume themselves and society in fruitless struggle, a power, apparently standing above society, has become necessary to moderate the conflict and keep it within the bounds of 'order;' and this power, arisen out of society but placing itself above it and increasingly alienating itself from it, is the state.[26]

To make this power more accessible, visible or 'real', the state is endowed with the most basic attribute of objects – location and extension in space. In civilisation, political power of the state is areal or territorial. The state is reified by placing it in space. Territorialisation of authority provides an open-ended assertion, and, if successful, exertion of control. By expressing power territorially there does not need to be a complete specification of the objects, events and relationships, which are subject to the authority of the state. Anything, both known and unknown, can fall under its authority if located within its territory. This open-ended means of asserting control is essential for the political activities of civilisations because it involves, almost by definition, the confrontation of novel and the unforeseen events; and such an unspecified domain can be claimed only through areal or territorial authority. Whatever else a state may be or do, it is territorial.[27]

The linking of society to place is more of a conscious effort in civilisations than it is in primitive societies and its function in the former is more clearly to reify a power and authority which, because of vastness and complexity of civilisations, is not clear and self-evident. Consciously moulding society into a territory tends to place more emphasis on the *territorial definition of society* than on the *social definition of territory*. The former has meant that social relationships are determined by location in a territory primarily and not by prior social connections, whereas the latter has meant that the use of an area or territory depends first and foremost on belonging to a group (the determination of which is essentially non-territorial).[28] In civilisation, a person's domicile frequently determines the person's membership in social organisations. Each location may be part of several overlapping or hierarchical jurisdictions so that being a resident of a place often means being part of several communities.

While a more territorial definition of power is a characteristic of civilisation, the degree to which all members of a society are aware of this varies tremendously. No civilisation has attempted to make its citizenry more aware of the central authority and the territorial extent of that authority than the modern nation state. Yet, many states in the twentieth century contain tribal and peasant cultures which are only minimally affected by modern conceptions of political territoriality and land use.

In addition to containing tribal cultures, a large proportion of the population of developing states contains peasant societies whose conceptions of place often combine elements of both primitive and modern. The peasant's conception of place focuses on the village and its surroundings. His relationship to the land is extremely close, personal and often mystical. Yet, his livelihood depends in part on contact with the city and the government, and from such contacts, some 'modern' conceptions have been incorporated within the more traditional and inward-looking views of peasant communities.[29]

A far greater mixture of civilised and primitive existed in the ancient states. Civilisations arose slowly and only partially replaced primitive societies. Often, the two societies lived side by side, or primitive forms of community existed on the local and rural levels within the territorial limits of the state. Much of the detail of the original transformations from primitive to civilised is lost and we have only a general idea of the process in some places. In these cases, there is also evidence of the accompanying changes in the association of society to place.

In ancient Egypt, for example, the original tribal or clan territorial units were called *spats* or *nomes*. As Egypt progressed from a tribal to a centralised empire, the *spat* remained a basic territorial unit but its relationship to the people changed. These units were no longer the demarcation of tribal holdings, but instead became an administrative area or province of the empire; and the position of leaders of the *spat* changed from the chiefs of the older tribal communities to administrators or governors of the province.[30]

For Greece and especially Attica, there exists documentation of the end of the transition from tribal to civilised society. At the beginning of the historical record, the Athenians still had vestiges of tribal structure. In the Heroic age they were composed of four tribes which were settled in separate territories. The tribes were composed of phratries and clans and the government of the Athenians was a tribal council or *Basileus*.

As the society of Athens became more complex, its inhabitants became more intermixed geographically but more clearly demarcated into classes, with the political and economic power being concentrated in the hands of the few in the upper class. The social territorial changes were expressed by the introduction of such institutions as the *Naukrariai*. The *Naukrariai*, established some time before Solon, were 'small territorial districts, twelve to each tribe', which were to provide and equip a warship and horsemen for the city of Athens.[31] According to Engels, this institution was one of the earliest recorded examples of the shift to a territorial definition of society. It attacked the older tribal form of association in two ways. First, 'it created a public force which was now no longer simply identical with the whole body of the armed people; secondly, for the first time it divided the people for public purposes, not by group of kinship, but by *common place of residence*'.[32]

Cliesthenes established a new constitution which completely ignored the older tribal territories. The new order was based on a territorial organisation of society. 'Not the people, but the territory was now divided: the inhabitants became a mere political appendage of the territory.'[33] Yet, to maintain sentiment to place the new territorial definition of society had to appeal to the older social definition of territory. In this regard, to keep sentiment to place alive, the new territory was a 'local' tribe. But the territory or local tribe was an artificial geographic unit for the convenience of the state. It was not an area which traditionally belonged to a group of related people as in the older tribes which were abolished. The basis of the territorial organisation were approximately 100 districts called *demes*. Residents of each *deme* elected a president, treasurer and judges. To again tap the older tribal attachments to place, the *demes* had their own temple and patron or divinity. The *demes* were organised into thirty *trittys* of approximately equal population, and from these *trittys* were formed ten 'local tribes'. The local tribe raised arms and men for the military and elected fifty representatives to the Athenian council which was composed of 500 representatives; fifty for each of the ten local tribes.

Such changes in the relationships between society and place can be seen in other areas of the world. Civilisation, with its emphasis on the territorial definition of society, makes social order closely bound to place. Community membership is often decided by domicile, and territorial defence (unlike the vagueness of the concept in tribal societies) becomes a primary obligation of the state. An attack on territory is a challenge to the state's order and authority.

It is the social complexity, inequality and the need for control of one group by another which make the territorial definition of society essential in civilisation. Paradoxically, it is the same forces which make the fusion of society to place in civilisation far more tentative and unstable than in the primitive world. On the one hand, territories constrain flows and movements of goods. On the other hand, separating activities in places creates specialisation which in turn increases the demand for trade and circulation. Activities which are readily located or contained within the territorial boundaries come to be thought of at the societal level as place specific, territorial or, in general, as spatial activities, whereas the flows and movements which are not readily contained come to be thought of as the less spatial or non-spatial activities.

Moreover, on the territorial or spatial side there develop degrees of 'spatiality'. Civilisations have several kinds of territorial units existing simultaneously. Modern nation states, for example, are divided into several sub-jurisdictions and these in turn may be divided into lower order districts creating areal hierarchies of territorial units whose actions may often be uncoordinated and in conflict with one another. A hierarchy of jurisdictions, or even a national territory divided into a single level of lower units, can create circumstances which are thought to be more or less spatial. A policy which is formulated at the national level may have predictable spatial consequences at that level, but perhaps not at the lower level of the geographic scale. Hence the spatial consequences of an action become foreseen at only one artificial scale. From the viewpoint of those who want to know the spatial consequences at the lower levels, or at all levels, the policy is not specific enough, and hence 'less' spatial than was desired.

The kinds of activities which come to be seen as less or non-spatial and the degree to which these conceptions are held by the citizens depend on the dynamics of the societies and on the ways in which the societies are anchored to place. In each society, the 'spatial', 'non-spatial' distinctions can appear at various levels of abstraction, from the material level of institutions and economics to the level of political and social philosophy.

Material and institutional separations of space and substance

The specific fusion of society to place in Western Feudalism differed from the fusion in the Graeco–Roman civilisations and in our own. Feudalism began as the absorption of a disintegrating Roman Empire by the Germanic tribes; it was an absorption and accommodation of a civilised view of territory by a primitive one. The Roman institutions of *precarium*, which was a dependent form of land holding, and the *petrocinium*, which was the offering of one's services for protection, were to form the basis of the fiefs and the vassalages of the feudal system. A peasant swore allegiance to a lord, and promised to meet such obligations as paying taxes on crops and rendering specific services to the lord, and in return the lord was to provide protection for the peasant. The entire structure rested on the labour of the peasant who did not own the land on which he lived. The peasant's lord owned the land but only in degree, for he too received the land with obligations from a noble of a higher order, and so on, up the social scale until the highest level – the king. In theory, there was to be no land without a sovereign. *Nulle terre sans seigneur.*

The peasants were bound to the earth, *glebae adscripti.* As long as the peasant fulfilled his part of the obligation he could not in theory be removed from the land. However, the peasant could not choose to leave the land either. While a lord could not sell a peasant as one could a slave, he could exchange land with another lord and, as a result of the exchange, would come a new lord for the peasant. For the peasants, social obligations and relations at the societal level could be determined by domicile. The survival of some communal village land and peasant holding from pre-feudal times did not appreciably weaken the peasant's bondage to the land.

Towns existed in the interstices of the manorial system. While towns were usually modest in size, they were important elements of the medieval economy and formed a fusion of place and society which was coexistent but in many respects contrary to the manorial system. A different form of law pertained within the territorial area of the city. City air, as the German saying went, makes man free. These special laws originated from the need of the merchants to have a permanent place of trade in the towns so that they could set up their wares and stay during inclement seasons. The inhabitants of these special places needed the right to travel, and those who sought the merchants' goods needed to be free to come and go. Such liberties of movement were granted, and they were paradoxically place specific. 'Freedom became the legal status of the bourgeoisie, so much so that,' according to Pirenne, 'it was no longer a personal privilege only but a territorial one, inherent in urban soil just as serfdom was in manorial soil. In order to obtain it, it was enough to have resided for a year and a day within the walls of the town.'[34]

Craft guilds formed another important aspect of the territorial authority of towns. These guilds established a craft monopoly within the towns. They participated in the government of the community and it was difficult for a person to engage in manufacture or become a labourer in towns without being a member of that community's guilds. These guilds were for the most part not areally interrelated (mercantile guilds were). Each tended to operate within the bounds of the city.

Both city and manor formed different, yet contemporaneous, territorial social organisations. The territorial unit of the city expanded in influence and helped transform the manorial system to commercial agriculture. The domination of the city occurred in part because of the increased importance of such activities as the flow of goods, people and money which were not containable within the existing territorial boundaries. Such flows which were not containable by the manorial system later became the basis of the less or non-spatial activities in industrialised societies. Even in the Middle Ages, such activities were not accorded equal status with property in land as a real thing. Property in land was, and is still, called *real estate*. The other is called *liquid assets*.

Although the coexistence of such territorial units as cities and manorial systems engendered processes which were incapable of territorial confinement or expression, the primary 'a-spatial' factor at the societal level was the idea of the Christian community. Indeed, the

Church, as an organisation, was spatial. It was territorially administered and appeared on the ground in the form of churches and holy places. These territorial aspects often conflicted with the territorial units of the secular society and also with the more ethereal concerns of the Church. But from the societal view, the primary opposition which the Church presented to the feudal order was the concept of an 'a-spatial' community of Christians. This community or heavenly city transcended terrestrial communities and boundaries. Unlike the 'non-spatial' aspects of society in the contemporary world, the Christian community of the church transcendent was associated with the fixed and the eternal, while the earthly cities were short-lived and changing.

The idea of a transcendent Christian community pervaded Christendom, had an enormous impact on feudal society, and contributed to the later conflicts between church and state, in which the power of the state triumphed. It was the rise of capitalism, however, which most fundamentally contributed to territorial reorganisation within the state. With the rise of merchant and then industrial capital, the means of production became concentrated in the hands of capitalists and spatially concentrated in workshops and manufactories wherein the labourer worked for wages under the supervision of management. Work thus became separated from the home and territorial control became specialised to include work places (under the control of industry and supported by government through property laws etc.) and political territories. The new economic order needed large and mobile labour pools and 'free' trade with a reliable and efficient transportation infrastructure. All of these factors altered the older territorial fusion of society to place. Coordination of economic functions was achieved by shifting the basic fusion of society and place to the larger geographic scale of the absolute state and then to the modern nation state. This left the cities with modest powers, as one of several territorial units in an areal hierarchy of state territorial organisations. Inhabitants of a city were not only citizens of that city, but also citizens of higher and lower political administrative units in which their residences were located.

As we well know, the anchoring of society to place in the nation state, with its hierarchy of territorial units, has not prevented conflicts between types and levels of territorial organisations, nor contained what are thought to be the less spatial processes. If anything, the opposite is the case. As the efforts to contain and anchor the system become more complex and self-conscious, as more levels of administration are created, the activities and interrelations that involve the spatial and non-spatial distinctions increase enormously. Because of the areal hierarchy of the political administrative system, the spatial manifestations of decisions are fragmented. Decisions may be directed to only one level of the hierarchy and yet affect all levels in unforeseen ways. Jurisdictional conflicts arise concerning the territorial units and the actions of one has unforeseen consequences on the others.

Antitheses between 'spatial' and 'non-spatial' facts or activities can be seen in the economic realm as well. In capitalist societies, parcels of land, clearly demarcated in place, are privately owned and held for purposes of speculation. The land may have been purchased because of its potential value, based on what might happen on or near it. Although the land contains substances, such as soil and vegetation, and perhaps even social factors like low-income families, the value of the land to the speculator may be determined solely by the future activities that could occur near or on it. A new highway may be constructed nearby which would increase the value of the property as a commercial site. Even after the highway is built, the owner of the land may not sell or build until the price is right. In such cases, the economic system makes us think of land as though it were empty, void of substances that have value, and of substances as though they were a-spatial entities existing abstractly somewhere but not materially on the land. Only under profitable circumstances do particular places and substances combine, and then only until a more profitable arrangement appears to disassociate and recombine them with other places and things.[35]

Philosophical separations of space and substance

The conceptual separation and attempts at recombination of society and space are most sharply illustrated in the realm of political philosophy. Philosophical speculation about social order is, as Diamond suggested, an outcome of societies which find such order difficult to attain. Such speculation is therefore largely absent in primitive societies. Attempts at social philosophy abstract the social system and conceptually separate space and social substances or activities. But not every attempt has discussed their recombination in detail or at all. This is especially the case in modern philosophy where discussions of specific parts of the social system such as education or justice occur with complete disregard for the possible spatial contexts of the society, and where spatial units can be discussed without regard for the wide range of social objectives which such units would be expected to help achieve. Such cases of leaving society relatively abstracted from space, or of space from society, point out the degree to which intellectual thought about society has become compartmentalised into concern about its spatial and less or non-spatial aspects.

But in the long history of speculations about social order, there have been extensive and influential examples of attempts to link society and space. These links are so complex that they often need to draw on several modes of thought, the combinations and emphases imparting to these attempts their particular philosophical flavours. Some, such as Bacon's *The New Atlantes*, are attempts at using science and technology primarily to portray the perfect habitat. Others, such as Orwell's *1984*, are caricatures of such attempts. Others, such as Campanella's *The City of the Sun*, rely primarily on a return to mythical consciousness to achieve the perfect order.

Perhaps the most comprehensive and influential attempts to link space and society are those of Plato and Aristotle. Greek city states at that time were classic examples of civilisation, but surrounding them were despotic empires and barbarians. The presence and perhaps the recollection of the primitive seemed so strong that in Plato's *Republic* there is a deliberate attempt to exclude all those justifications of the state which are based on the mythical–magical and the irrational. This is seen in Plato's proposals to control education, the arts, drama and poetry. But it is most poignantly demonstrated in his views about the initiation of the young into the Republic. Kinship ties, as we have noted, are fundamental bonds in primitive society. In the Republic, Plato would have the first generation sever any and all connections with family. This would be accomplished by

sending out into the country all the inhabitants of the city who are more than ten years old, and [the philosopher kings] will take possession of their children, who will be unaffected by the habits of their parents; these they will train in their own habits and laws, I mean the laws which we have given them: and in this way the State and constitution of which we were speaking will soonest and most

easily attain happiness, and the nation which has such a constitution will gain most.[36]

Once these ties are severed and the populace denied access to other
than sanctioned knowledge, the business of running a perfectly just
state could get underway.

To place this ideal society in a geographical context, Plato relies on
very rational and plausible suppositions about human behaviour
(one might say the equivalent then of social science generalisations).
We find these suppositions, for instance, in his appeals for mode-
ration on the size of the state.[37] However, when it comes to the
extremely important problem of instilling loyalty to the state, Plato
abandons the rational. He would have the loyalty of the people
directed to the most real and visible aspect of state authority, the
territory; and territorial allegiance was to be instilled by employing
the mode of thought which most effectively binds sentiment to a
specific place, the mythical. The myth which Plato introduces
somewhat sheepishly is modelled after many of the myths of creation
in which a people are born in a particular place from the womb of the
earth. 'I really know not how to look you in the face,' says Plato,
through Socrates

> or in what words to utter the audacious fiction, which I propose to
> communicate gradually, first to the rulers, then to the soldiers, and
> lastly to the people. They are to be told that their youth was a
> dream, and the education and training they received from us, an
> appearance only; in reality during all that time they were being
> formed and fed in the womb of the earth, where they themselves
> and their arms and appurtenances were manufactured; when they
> were completed, the earth, their mother, sent them up; and so, their
> country being their mother and also their nurse, they are bound to
> advise for her good, and to defend her against attacks, and her
> citizens they are to regard as children of the earth and their own
> brothers.[38]

Questions about the relationships between society and space,
about the ideal location and spatial configuration of activities, were
to be addressed throughout the history of social philosophy. What
changes there have been in such analyses concern the answers and the
mixtures of modes of thought from which the arguments are drawn.
Most of the current attention in political–social theory concerning

the relationships between society and place has been given to the scientific mode of expression, that is to arguments employing or implicitly appealing to plausible generalisations or explanation sketches about human behaviour. Because of the increased complexity and size of nation states and the fragmentation of societal thought into specific disciplines, there have been fewer attempts at exploring the relationships between spatial relations and the entire society, and more of a concern with how a particular spatial configuration would affect a particular aspect of behaviour. Because of the lack of a well-developed social science, such explorations, whether broad or narrow, have always resulted in partial, often inconclusive and contradictory answers.

For instance, the relationship between size and efficiency has recently been explored by using several different criteria for both. If we define efficiency as delivering services to a population at the lowest cost, we then may wish to determine optimal sizes for communities receiving these services. Communities smaller than the optimal are inefficient because they have not reached their economies of scale, and communities larger than the optimal are inefficient because they have diseconomies of scale. The optimal sizes vary from service to service. If, however, we consider the coordination of services as part of the meaning of efficiency, optimisation of units by size may run counter to optimising according to coordination. The former may lead to many independent and areally non-accordant jurisdictions while the latter may lead to few and more accordant ones. So far these considerations are primarily conjectural for there have been few significant measures of efficiency and finding a balance between the two approaches seems a long way off.

To complicate matters further there are other conditions of efficiency which may not be commensurable with the ones above. Too many areally non-accordant jurisdictions may decrease efficiency by making it difficult or impossible for citizens receiving the services to know who is responsible in case there is something wrong with the delivery of these services. But the coordination of services into large units may decrease efficiency, because it could reduce both the ability of a citizen to affect the outcome of political decisions and a citizen's sense of his own political efficacy. Restricting the number of areas and mixtures of services in terms of quantity and quality also limits the variety of environments from which people can select when they make residential choices.[39]

Problems of coordinating space and behaviour affect even political representation. Systems of representation are always less than perfect and areal representation introduces its own imperfections. Gerrymandering (the drawing of political boundaries to include a homogeneous electorate) was a familiar device which took advantage of such imperfections to help predetermine the outcome of elections. Even with a commitment to one-man, one-vote, and a policy of redistricting as mandated by the Supreme Court, areal representation still presents conflicts between domicile and fair representation. Some of these problems have been addressed by such devices as *at large representation, proportional representation* and *weighted multi-member districts.*[40] But each of these also has its own drawbacks to fair representation.[41]

There are many other areas at the societal level in which space and substance are conceptually separated, and need to be combined and interrelated in an understandable and predictable way. To discover the interrelationships we must determine the significant substance referents of the areal units, and this is an extremely difficult task. All kinds of substances and interactions can occur in an area, and unless these and their spatial properties are known, the 'area' and its spatial properties will not be sufficiently well specified in terms of substance referents. We have seen how generalisations about the distances between people in normal conversations must take into account substances through which the interactions are transmitted. Similarly, generalisations about political and social communities' propinquity, and overcrowding must be based on a knowledge of the kinds of interactions and the means of transmission that occur among individuals. There are some communities, such as families, which need physical contact, and others, such as a community of scholars, which can exist by the transmission of ideas. Similarly, significant measures of density depend on the activities undertaken. A mass of people packed densely together to watch a football game is not overcrowded, while a person who seeks solitude in a field may find the presence of only one other an intrusion.[42] Knowing the significant substance referents of the spatial terms means knowing the critical aspects of behaviour and this is the same as discovering social science generalisations.

Hybrid modes: planning

Understanding the spatial manifestations of behaviour becomes increasingly urgent as social relations continue to increase in complexity. There are several other modes of thought which can help connect space and behaviour and which have been employed on both the practical and the philosophical levels to affect the multiple objectives of society. But the preferred mode for integrating space and substance on the conceptual level in the twentieth century is the scientific. The social sciences are not strongly scientific, yet they have had an enormous impact on the way problems are approached. One would expect that if there were laws of human behaviour, social planning, like engineering, would be the application of laws to particular problems. But the social sciences, or perhaps more precisely the ideas or ideologies of the social sciences, nevertheless have had an enormous impact on our society through what we will term *scientific planning*. This type of planning often incorrectly uses incomplete generalisations of social sciences and also relies on 'procedures' of science which are mistaken for the essence of science.

Of those generalisations in the social sciences which have been confirmed or supported experimentally, most require extensive *ceteris paribus* assumptions. Such generalisations, though, are often incorporated into the plans of a community or used to support community goals, wherein the applications involve new and different contexts from the ones which were involved in the testing of the generalisations. This is an inappropriate use of generalisations for it is difficult to know if such generalisations will hold in these new contexts, and if not, by how much. The idea of a social science and its impact on society also involves the mimicking of so-called procedures or methods of science – namely quantification and deduction, and using these procedures in lieu of scientific generalisations. In scientific planning, rigorous and quantifiable deductions are often made from a set of assumptions which themselves are largely untested, perhaps plausible, but which may have no foundation in terms of a theory of behaviour. Furthermore, the deductions are not made to be tested; they are made to be implemented. By using the image and not the substance of science, scientific planning is a hybrid mode of science, peculiar to the scientifically oriented societies.

The use of incomplete generalisations, the inappropriate use of generalisations, and the substitution of deduction and quantification of scientific relationships and generalisations all come about in part because there is such an overpowering belief in a science of man as a way of solving problems of human behaviour that even the image of a science, without substance, carries weight. The positive aspects of this type of planning are that at least we can make our assumptions and deductions explicit. The negative aspects are that no matter how explicit may be our assumptions and rigorous our deductions, without scientific generalisations we still will not know the consequences of our plans or actions. Because the idea of science lends authority to any enterprise which involves the title science, or some of the procedures of science, many people, including planners, may not be aware of the fact that using scientific planning is not using science.

We see cases of scientific planning at the geographic scale in the use of Central Place assumptions in the planning of new towns. For example, the principles of regular spacing and areal and functional hierarchies were used in the development of new towns in the reclaimed lands in the Netherlands and in the development of new towns in southern Israel.[43]

The Netherlands example, perhaps more than any other, approximates to some of the conditions mentioned in Central Place Theory such as an isotropic plane, a relatively homogeneous agricultural population and the existence of towns providing primarily service functions. Even though these conditions were met approximately, and the settlements spaced rationally, the actual behaviour of the people did not correspond to what was planned. Some centres increased in population beyond what was expected and some decreased, and the mechanisation of agriculture progressed at a greater rate than was anticipated. In the Israeli example, the population was not homogeneous and the landscape not as isotropic as was the case in the Netherlands. The application of principles of spacing and hierarchy were therefore tempered and the plans were adjusted to include other factors. Despite these adjustments, the behaviour of the inhabitants did not accord with the expectations.

It is difficult, however, to assess the degree to which the actual behaviour does not coincide with the predicted in these and any other cases of planning, for to do so we would also need to know what the behaviour would have been if other plans or no plans at all were used. To know this is, of course, to have a theory of human behaviour. Moreover, in attempting to assess the impact of plans we should also be aware of the complications arising from the fact that a planned environment, especially one that has been designed or sanctioned by scientific planning, may tend to become a self-fulfilling prophecy. Simply by planning an environment we restrict the number of actions the inhabitants can make and if the inhabitants believe there was good reason for these restrictions, their behaviour may then come to accord with the restrictions of the plans. Therefore, by using scientific planning principles we may arrive at the position of having our actions mimic our conceptions of scientific behaviour long before we actually have a science of human behaviour.

Because of its emphasis on measurement and deduction, scientific planning may exaggerate the significance of easily measurable properties. This is especially true of spatial relations. Planning often attributes a special significance to distances and shapes, and in so doing treats these variables in a non-relational or incomplete relational way. This emphasis is reinforced by the tendency of social sciences to consider non-congruent spatial relations as occurring on an isotropic plane. Hence, many planning schemes attempt to minimise distances, to plan according to symmetrical spatial arrangements, when such configurations may not in fact produce the desired results, or may produce unexpected ones. Park benches may be symmetrically aligned along park paths without consideration for view or for the possibility of conversation; furniture in hospital wards may be arranged for the convenience of the staff and not for the convenience of the patients; and old, low-income neighbourhoods, because of their high density, disrepair and isolation, may be thought to have no redeeming features and may be demolished, when in fact such environments may foster strong senses of community and place.[44]

Much of what is being mimicked in 'scientific' planning is the feelings of certainty that come from logical deductions and mathematical relationships. This feeling, rather than the factivity of the relationship, directs our behaviour. And since, as Langer has pointed out, this kind of feeling is so specific that we hardly recognise it as feeling, we are all the more surprised to find that landscapes planned according to such principles produce particular kinds of feeling – ones that are repetitious, ordered and symmetrical. Whether such feelings are desirable or not depends on our preferences and predilections. Some may find these landscapes reassuring, clean and comfortable; others barren and monotonous.

Because such landscapes are not based on a true science of human behaviour, their consequences are often unforeseen, unpredictable, hazardous and uncoordinated with landscapes planned according to principles from other modes of thought, thus creating new problems for planners and social scientists to resolve. Landscapes developed to capture our feelings about science are not sufficient to satisfy our needs even in the twentieth century, an age which tends to belittle or ignore other feelings. We are not so completely immersed in mimicking science that we ignore the facts and uncritically accept the imitation. We often recognise that the consequences of our plans are not what we expected them to be. This realisation prevents scientific planning from fulfilling the function of myth and magic. But in the absence of any other agreed upon mode, and in the absence of a science of human behaviour, scientific planning does play an important role in our efforts to conceptualise and solve problems of order.

The maintenance of social order in twentieth century society cannot rely entirely on the sophisticated but tentative links which social science and scientific planning provide. In actual practice, the fusion of society and place is accomplished through the combination of sophisticated and unsophisticated modes. The state uses the sophisticated modes of social science, it uses the arts and it taps the unsophisticated modes to anchor behaviour to place. The United States has a 'heartland', it has national monuments, shrines and 'holy' places such as the Capital, the Lincoln Memorial, Grant's Tomb; it has a mythologised past in the rugged frontiersmen and the 'noble savage'. Attachment to nation states in fact may be one of the clearest expressions of mythical–magical consciousness of place in the twentieth century. As Yi-Fu Tuan points out, sacred space tends to be a locus of power; it is clearly demarcated and set apart; it is supposed to be complete; and it demands the ultimate sacrifice for its defence. All of these characteristics apply to the modern nation state.[45] Each state, though, has different conceptual separations of people and place and depends upon different mixtures of the sophisticated and unsophisticated to recombine them. Understanding these differences is the spatial perspective to social dynamics. What marks the modern conception of societal space from the primitive is the range of mixtures available to the modern society, and the fact that the modern, unlike the primitive, has access to more specialised sophisticated views. With this access and the complexities of social life which engendered such modes, comes a high degree of uncertainty, detachment and scepticism about the significant relationships between social order and geographic area.[46]

Part V
Conclusion

8
Geography and Social Science

We began the last chapter with the analysis of the unsophisticated – fused societal conception of space in the primitive world and we ended discussing the dilemmas of the relationship between the unsophisticated – fused and the sophisticated – fragmented conceptions which characterise nation states. Chapter 7 has brought us full circle in our conceptual surface. We have returned to the original problem of describing and reconciling different views of space but this time from the perspective of the group. Space is experienced from all of these views but modern man is often at a loss to know which one applies in which context and what are the significant interconnections among them because activities have become so very complex and knowledge so very compartmentalised that the relationships between space and substance, between subjectivity and objectivity, have been both practically and conceptually severed. It is no longer clear what are the geographic consequences of our actions. Even a science of human behaviour has not always been seen to deal with the relationship between space and substance as it would necessarily have to if it were indeed a science. The few efforts to study the concept of space for human behaviour mainly have explored space only within the narrow confines of practical behaviour in a Western context. It has fallen to geography to consider the full range of meanings of space and the nature of their syntheses as part of geography's general task of understanding the earth's surface as the home of man.

As a beginning to such a programme, we analysed the symbolic structures of the conceptions of space and their logical inter-relationships. Each person in each culture sees the world from different combinations of modes under different occasions. All normal human beings have the potential to see and to express themselves scientifically, practically, mythically – magically, artistically etc. Those modes are intrinsic to human thought. Our conceptual surface, based on a realist theory of science, provided the perspective to explore these modes and to see them as potentials of the human intellect. Such a characterisation of human activity is empirical. It is no different than saying that under proper conditions a match has the potential to burst into flames. Just as the match needs the right conditions for it to ignite, so too do human intellectual potentials need the proper conditions to flourish.

Socio-material conditions influence the contents of the modes of thought, their degrees of separation and development, but they do not affect their elemental structures. Whereas the content of science or art has changed over the centuries, the fact that an endeavour can be labelled scientific (or artistic) means that despite its particular content it shares a structural similarity or a family resemblance with other endeavours in this class. Even Marxist theory, which is most conscious of employing concepts whose meanings change when applied to different historical contexts, still maintains general definitions of terms and structural relationships for all social organisations.[1] It is the essential elements of the modes and their conceptions of space that we were after in this book. We see this identification as a necessary first step in geography's task of providing comprehensive examinations of the meanings of space as they pertain to our understanding of the earth's surface.

In the future we must systematically consider the socio-material context and its relationship to the conceptions of space. (These matters were only touched upon in the book.) To do so we must have a general understanding of how socio-economic structures affect the development, articulation and degree of separation of the modes of thought. But even before this we need to work out the kinds and functions of the spatial organisations of the political economic systems themselves. Much work has already been done on the location and interactions of parts of these systems (especially in capitalism) in the traditional vein of spatial analysis, but there is the equally, if not more, important consideration of territoriality (that is, the control of areas such as domiciles, work places and political areas, in order to control individuals and their relationships) which has not received much attention. In Chapter 7 we considered the conceptions of territory but did not systematically explore the kinds of territories and the functions they provide.

For example, it seems that territoriality in general has the potential to facilitate the control of people and their relationships. It offers a means of asserting control without specifying in detail what is being controlled (that is, definition by area rather than by kind). It reifies the control and hence makes it impersonal. By developing hierarchies of control it provides a means of 'dividing and conquering' because people in lower level territories have circumscribed knowledge and responsibility. This is compounded by allowing the geographic consequences of an action to be defined at a larger territorial unit than the one an individual may be involved with. These and other consequences require, then, that a bureaucratic system exist to organise the territories and their limited responsibilities. It remains to be seen in specific contexts which potentials are actualised and what new ones are developed. Such explorations would lay the foundation for a general understanding of the spatial organisations of socio-economic systems. We could then use this understanding to explore the effect of socio-economic structures (and their spatial organizations) on the conceptions of space. Again in this book we have only touched on some of these. For example, we can point to the function that schooling performs in developing and extending the last stage in Piaget's scheme, by removing people from the objects of thought and requiring them to use symbols (the written word itself has influence over symbol use); or conversely to the role of limited technology and the lack of schools and writing in developing symbolic forms that seem natural and universal and which can be conflated with the objects they represent. Or we can point to the role of capitalism in fragmenting modes of thought and understanding of spatial relationships and in making the physical science characterisation of space dominant in social science and practical life.

Characterising the modes and their interrelationships in general, exploring the spatial organisations of socio-economic systems, and examining the relationships of the latter with the former, constitute a systematic programme for understanding the meanings of space. These systematic studies must, however, be tested against the actual historical geographical contexts. We must see how a people and a culture use different mixtures of modes and develop hybrid ones. This in turn may make us modify, refine or expand our characteristics of space and see their limitations. The same applies to testing general statements about the function of spatial organisations in socio-economic systems and to general statements about the relationships between socio-economic systems and conceptions of space.

While testing general statements would help substantiate, correct or reformulate them, it would also provide a more complete picture of the geography of individual peoples, of what places and landscapes are like in a fuller sense. The possibility of developing these systematic programmes and the need to apply them to particular contexts in order to correct them and to learn more about places leads us to an extension of the traditional geographic dialectic between the generic and the specific or chorological. The extended form of the generic would refer to views in myth and magic, in art, in the practical mode, in the child's mode, as well as to conventional social science generalisations about the spatial organisations of society. The extended form of the specific or chorological analysis would involve studying one or more of these statements and their relationships in a personal or social context, not primarily to amend and refine the generic characterisations, though that may very well be a result of the application, but primarily to assist in the portrayal of a particular people's conception of place and landscape.

The extended form of the generic–specific dialectic requires a coherent and flexible framework of analysis. We believe that a realist theory of science has the potential to provide such a framework and we adopted it to present the major conceptions of space and their abstract interrelationships. As we pointed out, a realist theory, by recognising the empirical warrant of abstract and theoretical concepts and structures would consider formulations of Freud, of Marx of Lévi-Strauss or Lévy-Bruhl as scientific, if their empirical statements conform to the facts. Certainly they are excellent candidates for at least partial scientific analyses of human behaviour. They do point out underlying tendencies and structures which could influence events under special circumstances. But we have seen that the most general structural characteristic of science itself, its discursive symbolic form, presents extraordinary problems for developing more accurate and comprehensive social scientific theories because of its inability to capture the subjective and its symbolic forms.

Perhaps more of the structure of human behaviour can be elucidated by other kinds of formal symbolic systems (such as multivalued logics) but it is likely that greater success will be had if we allow some of our analyses to come closer to natural or ordinary language. Whether and how such analyses would still be scientific and whether they could be embraced by a realist theory remains to be worked out.[2] To attempt it we would need once again to consider the forms of analysis, synthesis and validity employed in chorology and historical explanations. Extending the range of science to the boundaries of ordinary language would reflect the longstanding view that much of social science is expressible in this language. Moreover, it would prepare us to see that much of our understanding of human behaviour will depend on analyses which do not rely entirely on well-formulated generalisations, but which nevertheless are thoughtful, knowledgeable, and which offer syntheses that are critical and plausible. These analyses would produce cycles of syntheses and critiques rather than cycles of laws and theories and would direct more of our attention to specific contexts. When combined with the search for generalisations, the result would be a healthy and realistic dialectic between generic and specific; each 'problematising' the other.[3]

Geography and history are the two fields of study that would make the greatest use of such a dialectical relationship. By employing a range of explanations, these disciplines could offer a vast scope for the study of man and present such studies in a broadly understandable context of space and time. This would allow us a greater command of our own subject-matter. It would bring us in touch with the variety of human experiences, feeling, emotions and their symbolic forms. It can in effect bring us closer to the centre of the conceptual surface. The centre is what the humanists say we have missed. The centre exists, however, only because the conceptual surface exists and this surface constitutes a scientific perspective on space.[4]

Our formulation of the meanings of space depends, of course, on how valid are the characterisations of the modes of thought. The ones we selected had much to say about the meaning of space. But they are controversial and incomplete and alternative characterisations exist. Employing them may well alter the details of the conceptual surface, the relative locations of the modes and the descriptions of the impediments facing the social sciences. But the major problems of the conceptual separation and recombination of space, substance, subjective and objective are fundamental in Western thought and are likely to remain.

Notes and References

Chapter 1: Space and Modes of Thought

1. Realism has a range of meanings. For positivistic realism see, for example, Gustav Bergman, *The Metaphysics of Logical Positivism* (Madison: University of Wisconsin Press, 1967). For transcendental realism see Roy Bhaskar, *A Realist Theory of Science* (New Jersey: Humanities Press, 1978). For a discussion of the impact of realism on the social sciences see Russell Keat and John Urry, *Social Theory as Science* (London and Boston: Routledge and Kegan Paul, 1975); and Roy Bhaskar 'On the Possibility of Social Science Knowledge and the Limits of Naturalism', *Journal for the Theory of Social Behavior*, vol. 8 (1978) pp. 1–28.

2. For discussions of structuralism and phenomenology in geography see Yi-Fu Tuan, 'Geography, Phenomenology, and the Study of Human Nature', *Canadian Geographer*, vol. 15 (1971) pp. 181–92; and 'Structuralism, Existentialism and Environmental Perception', *Environment and Behavior*, vol. 3 (1972) pp. 319–31. For a discussion of idealism and geography see Leonard Guelke, 'An Idealist Alternative in Human Geography', *Annals of the Association of American Geographers*, vol. 64 (1974) pp. 193–202. For Marxism in geography see David Harvey, *Social Justice and the City* (Baltimore: Johns Hopkins University Press, 1973). For fuzzy sets and many valued logics see Stephen Gale, 'On the Heterodoxy of Explanation: A Review of David Harvey's *Explanation in Geography*', *Geographical Analysis*, vol. 4 (1972) pp. 285–322; 'Inexactness, Fuzzy Sets, and the Foundations of Behavioral Geography', *Geographical Analysis*, vol. 4 (1972) pp. 337–49; and Gunnar Olsson, *Birds in Egg*, Michigan Geographical Publications no. 15 (Ann Arbor, Michigan: Department of Geography, 1975). For a general critique of the philosophy of social science and human geography see Derek Gregory, *Ideology, Science and Human Geography* (London: Hutchinson, 1978).

3. P. F. Strawson, *Individuals: An Essay in Descriptive Metaphysics* (New York: Anchor Books, Doubleday, 1963) argues that a spatial system is the framework we have used for individuation. Any other would be extremely strange and cumbersome. See also the description of the role of space in A. Michotte, *The Perception of Causality* (London: Methuen, 1963).

4. For a discussion of the concept of space in the physical sciences see Hans Reichenbach, *The Philosophy of Space and Time*, tr. Maria Reichenbach (New York: Dover, 1957); Bastian von Fraassen, *An Introduction to the Philosophy of Time and Space* (New York: Random House, 1970); and Graham Nerlich, *The Shape of Space* (Cambridge University Press, 1976).

5. Recent advances in physics have made it clear that there can be several 'accurate' descriptions of physical space and that the 'best' description is one which makes the relationships among space, time, and energy, matter, or substance (as expressed in the laws and theories of physics) the simplest and most fruitful.

6. Richard Gallagher, *Diseases that Plague Modern Man* (New York: Oceana Publications, 1969) p. 24.

7. Ibid., p. 17.

8. Gerald Pyle, 'The Diffusion of Cholera in the United States in the Nineteenth Century', *Geographical Analysis*, vol. 1 (1969) pp. 59–75.

9. Among the works on methods geared to spatial relations are Leslie King, *Statistical Analysis in Geography* (Englewood Cliffs, N. J.: Prentice-Hall, 1969); and Maurice Yeates, *An Introduction to Quantitative Analysis in Human Geography* (New York: McGraw Hill, 1974). For a discussion of causal relations in non-experimental designs see Herbert Blalock, Jr, *Causal Inferences in Nonexperimental Research* (Chapel Hill: University of North Carolina Press, 1964).

10. There are, for instance, the issues of spatial auto correlation, areal sampling, aspects of ecological correlation, point–pattern analysis and areal subdivision procedures.

11. An example of both the aggregation of data and the obfuscation of its geographical location is the publication of a survey of Wisconsin farmers in the 1920s. The data originally contained interviews with settlers and records of the physical characteristics of their farms. Instead of keeping track of these individual units and their locations in space, the data was aggregated and the published sources refer to coded areas which are not related to known geographic locations. See 'Economic Aspects of Land Settlement in the Cut-Over Region of the Great Lakes States', *United States Department of Agriculture*, Circular No. 160, Washington, D C (April 1931).

12. The search for correspondences between mental states and physical

states is the basis of the behavioural approach and will be discussed in more detail in Chapters 2 and 4.

13. These are issues of form and process. For a discussion of them see William Bunge, *Theoretical Geography* (Lund Studies in Geography, Series C, General and Mathematical Geography, No. 1. Lund: C. W. K. Gleerup).

14. This, at least, is the way it appears to the outsider. The adept though may believe that there are 'forces' or 'powers' traversing space and causing such interactions. From the evidence it seems that most people who believe in magic and myth do not pay much attention to these matters and act as though things affected each other at-a-distance. It should also be noted that action at-a-distance can occur in the physical realm as in the interpretation of gravitational attraction in the Newtonian framework, but that physics does not suggest that it occurs for the kinds of activities on the surface of the earth of interest to social scientists and geographers.

15. Pierre Deffontaines, *Géographie et Religiones* (Paris: Gallimard, 1948) p. 118. See also Agnes Hoernlé, 'The Social Organization of the Nama Hottentots of Southwest Africa', *American Anthropologist*, vol. 27 (1925) pp. 1–24 (especially pp. 15–17); and Pierre Kolbe, *Description du Cap de Bonne-Esperance* (Amsterdam: Jean Catuffe, 1741) vol. 1, pp. 329–38.

16. For the image of the circle in Elizabethan poetry see Marjorie Nicolson, *The Breaking of the Circle* (Evanston, Ill: Northwestern University Press, 1950); and for the image of the circle in geography see Yi-Fu Tuan, *The Hydrologic Cycle and the Wisdom of God* (Toronto: University of Toronto Press, 1968).

17. Francis Quarles, 'Emblem VI', in *Emblems, Divine and Moral* (London: William Tegg, 1886) p. 20

18. John Milton, *Paradise Lost*, Book vii, 11, pp. 225–31.

19. José and Miriam Argüelles, *Mandala* (Berkeley and London: Shambhala, 1972) p. 23.

20. Rudolf Arnheim, *Visual Thinking* (Berkeley: University of California Press, 1969) p. 231.

21. Paul Bohannan and Philip Curtin, *Africa and Africans* (New York: The Natural History Press, 1971) pp. 120–8, discuss the views of territory in African societies.

22. For a discussion of the different views of land ownership and the ensuing conflicts between Indians and Whites, see R. W. Bryant, *Land: Private Property, Public Control* (Montreal: Harvest House, 1972) chapters 1 and 2.

23. The most important discussions of the meanings that space and place can have within differing views of science are found in Richard Hartshorne, *The Nature of Geography: A Critical Survey of Current Thought in the*

Light of the Past (Lancaster: Association of American Geographers, 1939 and reprinted with corrections, 1961); and Fred Lukermann, 'Geography: de facto or de jure', *Journal of the Minnesota Academy of Science*, vol. 32 (1965) pp. 189–96. Considerations of space beyond the scientific are in David Lowenthal, 'Geography, Experience, and Imagination: Toward a Geographical Epistemology', *Annals, Association of American Geographers*, vol. 51 (1961) pp. 241–60; and Yi-Fu Tuan's *Topophilia* (Englewood Cliffs, NJ: Prentice-Hall, 1974), and *Space and Place* (Minneapolis: University of Minnesota Press, 1977). In sociology see Erik Cohen, 'Environmental Orientations', *Current Anthropology*, vol. 47 (1976) pp. 49–70. For works in landscape architecture and geography see J. B. Jackson, *Landscapes: Selected Writings of J. B. Jackson* (Massachusetts: University of Massachusetts Press, 1970); and the journal, *Landscape*. It is important to note that there have been attempts to expand the conception of time in geography. See, for example, the works by T. Hagerstrand; the volume of *Economic Geography*, vol. 53 (1977), much of which is addressed to Hagerstrand's ideas of a space–time viewpoint in geography, and Nigel Thrift, 'Time and Theory in Human Geography: Part I', in *Progress in Human Geography*, new series, vol. 1 (1977) pp. 65–101.

24. Susanne Langer, *Feeling and Form* (New York: Charles Scribner's, 1953) pp. 45–85.

25. Susanne Langer, *Philosophy in a New Key* (Cambridge: Harvard University Press, 1957) pp. 79–102.

26. Susanne Langer believes that modes of thought such as art will not be amenable to logical forms. Gunnar Olsson, *Birds in Egg* (Ann Arbor: Michigan Geographical Publication No. 15; Department of Geography, University of Michigan, 1975) believes that different logics or inference rules may go a long way in capturing different patterns of thought.

Chapter 2: Science and Subjectivity

1. The most comprehensive review of positivism by a geographer is in David Harvey, *Explanation in Geography* (London: Edward Arnold, 1969). See also Stephen Gale, 'On the Heterodoxy of Explanation: A review of David Harvey's "Explanation in Geography"', *Geographical Analysis*, vol. 4 (1972) pp. 285–322; and the reply by David Harvey, 'On Obfuscation in Geography: A Comment on Gale's Heterodoxy', *Geographical Analysis*, vol. 4 (1972) pp. 323–30. For more recent criticisms of positivism see Chapter 1, reference 1, and Harold Brown, *Perception, Theory and Commitment: The New Philosophy of Science* (University of Chicago Press, 1977).

2. May Brodbeck, 'Values and Social Science: Introduction', in May

Brodbeck (ed.), *Readings in the Philosophy of Social Sciences* (New York: Macmillan, 1968) p. 79.

3. The following types and definitions of meaning were adapted from May Brodbeck, 'Meaning and Action', in May Brodbeck (ed.), *Readings in the Philosophy of the Social Sciences*, pp. 58–78.

4. Ibid., footnote 3.

5. Two qualifications should be made regarding this matter. First, in the social sciences, the number of conditions that often would have to be included for a generalisation to hold, even approximately, are enormous and would make such a generalisation appropriate for only a small number of actual cases. These cases in effect may have proper names or be particular cultures in particular places. Often, instead of including these variables, the social sciences especially employ a *ceteris paribus* assumption which is equivalent in many respects to saying that the generalisation would hold in a place or environment 'like this'. Second, we can have a law about a single place in the sense that the attributes of the place are not discussed as instances of concepts but are simply related to locational coordinates or specific places. For example, we can have a law about rainfall and corn yields for Wisconsin. But such laws would have to be reformulated so that all of the terms are about concepts for them to be included in or explained by other laws and generalisations.

6. See the discussion of static generalisations in Gustav Bergmann, *Philosophy of Science* (Madison: University of Wisconsin Press, 1966) pp. 103 and 118.

7. Karl Popper, *The Logic of Scientific Discovery* (New York: Harper Torchbooks, 1965) p. 41, makes the point that we cannot verify generalisations, we can only disconfirm them. '*It must be possible for an empirical scientific system to be refuted by experience.*' (Italics are Popper's.)

8. On the asymmetry of explanation and prediction see Gunnar Olsson, *Birds in Egg* (Ann Arbor: Michigan Geographical Publication No. 15, Department of Geography, University of Michigan, 1975) pp. 58–9.

9. A classic review of the 'pros' and 'cons' of scientific analyses of human behaviour is found in Ernest Nagel, *The Structure of Science* (New York: Harcourt, Brace and World, 1961) pp. 447–85.

10. Brodbeck, *Readings in the Philosophy of the Social Sciences*.

11. There are many formulations which describe the mind–body problem and the translations between the two. For a view of these problems in geography see Pat Burnett, 'Behavioral Geography and the Philosophy of Mind', in Reginald Golledge and Gerard Rushton (eds), *Spatial Choice and Spatial Behavior* (Columbus: Ohio State University Press, 1976) pp. 23–48.

12. Susanne Langer, *Mind: An Essay on Human Feeling* (Baltimore: Johns Hopkins Press, 1967) vol. I, p. 30.

13. Ibid., p. 29.
14. Ibid., p. 30.
15. Ibid., p. 31.
16. Ibid., p. 31.
17. Langer, *Feeling and Form* (New York: Charles Scribner's, 1953) p. 40.
18. Langer, *Feeling and Form*, footnote 17, especially p. 212.
19. Langer, *Mind*, pp. 147–8.
20. William Moody, 'Gloucester Moors', in *Gloucester Moors and Other Poems* (Boston and New York: Houghton Mifflin, 1910).
21. For the distinction between representation and presentation see Susanne Langer, *Philosophy in a New Key* (Cambridge: Harvard University Press, 1957) chapter IV.

Chapter 3: Social Science and Objective Meanings of Space

1. A. d'Abro, *The Rise of the New Physics* (Dover Publications, 1951) vol. I, pp. 112–13; and Einstein's 'Forward to Max Jammer', *Concepts of Space* (New York: Harper and Brothers, 1969).
2. Jammer, footnote 1, pp. 195–6, referring to Einstein's comments p. XIV.
3. Adolf Grünbaum, *Philosophical Problems of Space and Time* (New York: Alfred A. Knopf, 1963) p. 421 and footnote 9, referring to Einstein's views. For a discussion of the current status of the view of absolute space, see Graham Nerlich, *The Shape of Space* (New York: Cambridge University Press, 1976).
4. Mary Hesse, 'Action at a Distance and Field Theory', in *The Encyclopedia of Philosophy* (New York: Macmillan, 1967) vol. I, p. 12.
5. While action at-a-distance is a possible interpretation of some physical science laws like the Newtonian Gravitational Laws, it has never been a popular idea. See A. d'Abro, *The Rise of the New Physics*, pp. 106–14.
6. Relational *concept* rather than *theory* is used in order to distinguish it from the relational theory. The relational concept pertains to social science while the relational theory pertains to physical science. It is interesting to note that the philosophy of social science has discussed at length the concept of time in social science laws but has not carefully examined the concept of space in social science laws or generalisations.
7. Spatial terms include physical geometric terms as well as references to place and size. Substance terms refer to both 'objects' and 'attributes'.
8. May Brodbeck, 'Methodological Individualisms: Definition and Reduction', in May Brodbeck (ed.), *Readings in the Philosophy of the Social Sciences* (New York: Macmillan, 1968) pp. 280–303, esp. pp. 284–6.
9. Because the social sciences usually discuss only two of the dimensions of space, the congruence of two facts can refer to their shared surfaces.
10. For a discussion of proxemics see Edward Hall, *The Hidden Dimension*

(Garden City, New York: Anchor Books, 1969); and for a discussion of geographic generalisations see Peter Haggett, Andrew Cliff and Allan Frey, *Locational Analysis in Human Geography*, second edition (London: Edward Arnold, 1977).

11. The laws of gravity in Newtonian physics can be interpreted from the framework of action at-a-distance. See footnote 5.

12. Fred K. Schaeffer, 'Exceptionalism in Geography: A Methodological Examination', *Annals, Association of American Geographers*, vol. 3 (1953) pp. 226–49; and William Bunge, *Theoretical Geography* (Lund Studies in Geography, Series C, General and Mathematical Geography, No. 1, C. W. K. Gleerup, 1966). It should be pointed out that space or distance can be used non-relationally as a first approximation if we recognise that it is just that, a first approximation, and that its applicability will be limited.

13. See the geographical or spatial inference problem as discussed in Gunnar Olsson, *Birds in Egg* (Ann Arbor: Department of Geography, University of Michigan, 1975) p. 465.

14. See the discussions of geography, history and the social sciences in Richard Hartshorne, *The Nature of Geography* (Lancaster, Pennsylvania: Association of American Geographers, 1939) *passim*.

15. See, for example, Arthur Maass (ed.), *Area and Power* (Glencoe, Illinois: Free Press, 1959) pp. 22–6 for a discussion of the underemphasis of the 'local level' in political science.

16. Alfred Marshall, *Principles of Economics* (London: Macmillan, 1936) 8th edition, Book v, chapter xv, section 1.

17. The reaction against the exclusion of space in economics has occasioned the rise of regional science which is closely related to economic geography. See Walter Isard, *Location and Space Economy* (Cambridge, Massachusetts: M.I.T. Press, 1965) and *Methods of Regional Analysis* (Cambridge, Massachusetts: M.I.T. Press, 1966).

18. Peter Gould, *Spatial Diffusion* (Commission on College Geography Resource Paper No. 4, 1969) p. 19.

19. Walter Isard, *Methods of Regional Analysis*, footnote 18, p. 190.

20. Edward Hall, *The Hidden Dimension* (New York: Doubleday, 1969) p. 160.

21. William Warntz, 'Geography of Prices and Spatial Interaction', *Papers and Proceedings, Regional Science Association*, vol. 3, (1957) p. 128.

22. J. Stewart, 'Discussion: Population Projection by Means of Income Potential Models', *Papers and Proceedings, Regional Science Association*, vol. 4 (1958) p. 153.

23. Maurice Yeates and Barry Garner, *The North American City* (New York: Harper and Row, 1971) p. 114. See also Gunnar Olsson, *Distance and Human Interaction* (Philadelphia: Regional Science Research Institute, 1964).

24. Yeates and Garner, *The North American City*, p. 103. We are referring here to a popular version of the 'Law'. The original by W. Reilly, 'Methods for the Study of Retail Relationships' (Bureau of Business Research Monograph 4, University of Texas, 1929) pp. 1–50 is more precise about substance referents. See especially pp. 16 and 18. It is also the case that the more recent formulations of Central Place Theory are less concerned about substance referents of the spatial terms than was the original by Walter Christaller, *Central Places in Southern Germany*, trans. by Carlisle Baskin (Englewood Cliffs, N J: Prentice-Hall, 1966).

25. S. A. Stouffer, 'Intervening Opportunities: A Theory Relating Mobility and Distance', *American Sociological Review*, vol. 5 (1940) pp. 845–67.

26. Reginald Golledge and Douglas Amedeo, An Introduction to Scientific Reasoning in Geography (New York: John Wiley, 1975) p. 216 and p. 220.

27. Ibid., pp. 299–300.

28. For one of the better attempts see Bunge, *Theoretical Geography*, pp. 73–80.

29. For academic interest in the theory see Ian Hodder and Clive Orton, *Spatial Analysis in Archaeology* (Cambridge University Press, 1976). Some of the ideas of the theory have been used to plan new settlements in the Netherlands and in Israel. See M. von Hulton, 'Planned Reality in the Ysselmeer-Polders', *Tydscrift Voor Economische an Sociale Geogràfie*, LX (1969) pp. 67–76; and Yehoshua Ben Arieh, 'The Lokhish Settlement Project', *Tydscrift Voor Economische an Sociale Geografie*, LXI (1970) pp. 334–47.

30. See the discussion of geographic explanation and laws in Schaeffer, 'Exceptionalism in Geography', and Bunge, *Theoretical Geography*.

31. See William Bunge, 'Spatial Prediction', *Annals, Association of American Geographers*, vol. 63 (1973) pp. 566–8; and Robert Sack, 'Comment in Reply', *Annals, Association of American Geographers*, vol. 63 (1973) pp. 568–9.

Chapter 4: Social Science and Subjective Meanings of Space

1. This is the conclusion Richard Hartshorne reaches in *The Nature of Geography* (Lancaster, Pennsylvania: Association of American Geographers, 1939). This book is the clearest and most influential statement of chorology in the English language. See also his *Perspective on the Nature of Geography* (Chicago: Rand McNally, 1959).

2. See the discussion of the generic in Richard Hartshorne, *The Nature of Geography*, passim.

3. See the discussion of the specific in Ibid., footnote 1, *passim*.

4. Ibid., footnote 1, p. 415.

5. The idea of 'local conditionality' is discussed by Fred Lukermann, 'Geography: de facto or de jure', *Journal of the Minnesota Academy of Science*, vol. 32 (1965) p. 194.
6. Carl Hempel, 'The Function of General Laws in History', *Journal of Philosophy*, vol. 39 (1942) p. 42. For recent amendments to the idea of explanation sketches see Terry Goode, 'Explanation, Expansion, and the Aims of Historians: Toward an Alternative Account of Historical Explanation', *Philosophy of the Social Sciences*, vol. 7 (1977) pp. 367–84.
7. Hempel, 'The Function of General Laws in History', p. 42.
8. Another factor contributing to the tentativeness of such discourses is their use, either implicitly or explicitly, of incomplete statistical relationships.
9. 'Behavioural' has many meanings. We are using the term here as it has been used in geography and this use is close to the philosophical position of 'logical behaviouralism'.
10. See Gustav Bergman, *Philosophy of Science* (Madison: University of Wisconsin Press, 1957) pp. 166; 170–1.
11. For a review of the behavioural approach in political science, see James Charlesworth (ed.), *Contemporary Political Analyses* (New York: Free Press, 1967).
12. Martin Cadwallader, 'A Methodological Examination of Cognitive Distance', in Wolfgang Preiser (ed.), *Environmental Design Research* (Stroudsburg, Pa.: Dowden, Hutchinson and Ross, 1973) pp. 193–9; and R. Day, 'Urban Distance Cognition: Review and Contributions', *Australian Geographer*, vol. 13 (1976) pp. 193–200.
13. O. Bratfisch, 'A Further Study of the Relation between Subjective Distance and Emotional Involvement', *Acta Psychologica*, vol. 29 (1969) pp. 244–55, and Ronald Briggs, 'Urban Cognitive Distance', in R. Downs and D. Stea, *Image and Environment* (Chicago: Aldine, 1973) pp. 193–9.
14. Martin Cadwallader, 'Problems in Cognitive Distance and their Implications for Cognitive Mapping', *Environment and Behaviour*, forthcoming.
15. The range of possibilities is discussed in Thomas Saarinen, *Environmental Planning* (Boston: Houghton Mifflin, 1976). We are using the term cognitive or mental map to refer to attempts to reconstruct a person's knowledge of the spatial relations of things on the earth's surface. The term mental maps especially has also been used to refer to the mapping of attitudes, values and preferences without concern for the person's knowledge of location and spatial configuration. In this regard see Peter Gould and Rodney White, *Mental Maps* (Baltimore: Penguin, 1974).
16. Waldo Tobler, 'The Geometry of Mental Maps', in Reginald Golledge

and Gerard Rushton (eds), *Spatial Choice and Spatial Behavior* (Columbus: Ohio State University Press, 1976) pp. 69–81.

17. Martin Cadwallader, 'A Behavioral Model of Consumer Spatial Decision Making', *Economic Geography*, vol. 51 (1975) pp. 339–49.

18. Terrence Lee, 'Psychology and Living Space', in Roger Downs and David Stea (eds), *Image and Environment* (Chicago: Aldine, 1973) p. 98; and Roger Downs and David Stea, 'Cognitive Maps and Spatial Behavior: Process and Products', in Roger Downs and David Stea (eds), *Image and Environment* (Chicago: Aldine, 1973) p. 11.

19. Martin Cadwallader, 'Problems in Cognitive Distance and their Implications for Cognitive Mapping'.

20. Yi-Fu Tuan, 'Image and Mental Maps', *Annals, Association of American Geographers*, vol. 65 (1975) pp. 205–12; and M. Merleau-Ponty, *Phenomenology of Perception* (London: Routledge and Kegan Paul, 1962) *passim*.

21. Yi-Fu Tuan, 'Image and Mental Maps'.

22. Ezekiel v: 5, *Holy Bible*, Revised Standard Version (1952).

23. Norman Thrower, *Maps and Man* (Englewood Cliffs, NJ: Prentice-Hall, 1972) pp. 32–4. See also Lloyd A. Brown, *The Story of Maps* (New York: Bonanza Books, 1949) pp. 94–100. For a general discussion and review of geographic views during the Middle Ages, see J. K. Wright, *Geographic Lore of the Time of the Crusades: A Study in the History of Medieval Science and Tradition in Western Europe* (New York: Dover Publications, 1965).

24. Isaiah XI: 12, *Holy Bible*, Revised Standard Version (1952).

25. See, for example, Charles Mahaffey, *Changing Images of the Cutover: An Historical Geography of Resource Utilization in the Lake Superior Region, 1845–1930* (unpublished Ph.D Dissertation, University of Wisconsin, Madison, 1978).

26. Hildegard Johnson, *Order Upon the Land* (New York: Oxford University Press, 1976).

27. For a discussion of Utopias from a geographic viewpoint, see Philip Porter and Fred Lukermann, 'The Geography of Utopia', in David Lowenthal and Martyn Bowden (eds), *Geographies of the Mind* (New York: Oxford University Press, 1976) pp. 197–223.

28. The impact of aesthetics on landscape is an extremely important source for knowledge about our feelings for shapes and patterns. For excellent analyses of the impact of aesthetics on landscape at the geographic scale see John Barrell, *The Idea of Landscape and the Sense of Place 1730–1840: An Approach to the Poetry of John Clare* (Cambridge University Press, 1972) and Kenneth Olwig, *The Morphology of a Symbolic Landscape: A Geosophical Case Study of the Transformation of Denmark's Jutland Heaths Circa 1750–1950* (unpublished Ph.D. thesis, Department of Geography, University of Minnesota, 1977). A dis-

cussion of *tastes* and their reflections in the landscape is found in David Lowenthal and Hugh Prince, 'The English Landscape', *The Geographical Review*, 54 (1964) pp. 309–46.

29. Sigmund Freud, *The Interpretation of Dreams*, Great Books of the Western World (Chicago: Encyclopedia Britannica, 1952) p. 255 (Freud's Italics.)
30. Ibid., pp. 265–277.
31. Ibid., p. 266.
32. Ibid., pp. 267–268.
33. Ibid., pp. 283–4. For a further discussion of symbolism in dreams see A. A. Brill, 'The Universality of Symbols', *The Psychoanalytic Review*, vol. 30 (1943) pp. 1–18, especially p. 4.
34. Carl Jung, 'Approaching the Unconscious', in Carl Jung *et al.* (eds), *Man and His Symbols* (New York: Dell, 1975) p. 58.
35. M. L. von Franz, 'The Process of Individuation', in Jung, 'Approaching the Unconscious', p. 234.
36. Yi-Fu Tuan, *Topophilia* (Englewood Cliffs, NJ: Prentice-Hall, 1974) discusses how diverse are our reactions to the similar landscape stimuli.
37. There have been some generalisations from art criticism which stand up to empirical validation as in those of Northrop Frye, *Anatomy of Criticism* (Princeton University Press, 1971) and there have been some generalisations from the realm of experimental aesthetics as in Daniel Berlyne (ed.), *Studies in the New Experimental Aesthetics* (Washington: Hemisphere Publishing, 1974). These generalisations are perhaps edifying but they are not counterintuitive or surprising. They do not show us what we may not already have known. Those works that have attempted to go beyond the most general levels and develop a series of generalisations, as with Joseph Schillinger, *The Mathematical Basis of the Arts* (New York: Philosophical Library, 1948), do not stand up to close scrutiny.
38. Susanne Langer, *Feeling and Form* (New York: Charles Scribner's, 1953) pp. 84–5.
39. Susanne Langer, *Mind* (Baltimore: The Johns Hopkins Press, 1967) vol. i, p. 90.
40. Ibid., p. 186.
41. Ibid., p. 157.
42. In this regard, art criticism is a form of social science. There are impressionistic forms of art criticism and art criticism that is itself art. Both of these, of course, are not part of scientific methodology.
43. William Wordsworth, *The Prelude* (1806–7 version) Book v, lines 56–139. This passage of the poem is analysed in W. Auden, *The Enchafed Flood* (New York: Random House, 1967). I am indebted to Ken Olwig for pointing this out.

Chapter 5: The Child's and the Practical View of Space

1. Heinz Werner, *Comparative Psychology of Mental Development* (New York: International Universities Press, 1973) pp. 40–56d.
2. Ibid., pp. 53ff and 111ff.
3. Ibid., p. 53, lists discrete, articulated, definite, flexible and stable.
4. Animism seems to be prevalent in children and occurs most frequently in the early stages of development. See ibid., pp. 72 and 80; Jean Piaget, *The Child's Conception of the World* (London: Routledge and Kegan Paul, 1971) pp. 169–252; and Bruno Bettelheim, *The Uses of Enchantment* (New York: Random House, 1977) pp. 45–7.
5. Werner, *Comparative Psychology of Mental Development*, pp. 64–5 quoting William Stern, *Psychology of Early Childhood* (German edition, 1930) p. 46.
6. Werner, *Comparative Psychology of Mental Development*, p. 65.
7. The works of Piaget are voluminous. In the following we have relied primarily on Piaget's, *The Child's Construction of Reality* (London: Routledge and Kegan Paul, 1955); *The Child's Conception of the World* (London: Routledge and Kegan Paul, 1971); *The Child's Conception of Space* (New York: W. W. Norton, 1967); *The Origins of Intelligence in Children* (New York: W. W. Norton, 1963); John Flavell, *The Developmental Psychology of Jean Piaget* (New York: P. Van Nostrand, 1963); and Alfred Baldwin, *Theories of Child Development* (New York: John Wiley, 1968).
8. The most succinct discussion of these relations is found in Flavell, *The Developmental Psychology of Jean Piaget*, p. 44.
9. The names and numbers of sub-stages vary in the references. We have combined Flavell's and Baldwin's taxonomy in the following. Flavell, *The Developmental Psychology of Jean Piaget*, and Baldwin, *Theories of Child Development*.
10. Piaget, *The Child's Construction of Reality*, p. 86.
11. Ibid., pp. 98–9.
12. Piaget, *The Child's Conception of Space*, pp. 6–8.
13. These four are taken from Jonas Langer, *Theories of Development* (New York: Holt, Rinehart and Winston, 1969) pp. 129–31.
14. Piaget, *The Child's Conception of Space*, p. 244.
15. Ibid., p. 375.
16. See, for example, the organisation of Piaget's *The Child's Construction of Reality*.
17. Piaget, *The Child's Conception of Space*, p. 454.
18. Ibid., pp. 448–9.
19. See, for instance, Rudolf Arnheim, *Visual Thinking* (Berkeley: University of California Press, 1974), and H. Price, *Thinking and Experience* (Cambridge: Harvard University Press, 1953).

20. The problems of perception are discussed in James Gibson, *The Senses Considered as Perceptual Systems* (Boston: Houghton Mifflin, 1966).
21. Yi-Fu Tuan, *Topophilia* (Englewood Cliffs, NJ: Prentice-Hall, 1974) p. 27. See also David Lowenthal's discussion, 'Geography, Experience and Imagination: Toward a Geographical Epistemology', *Annals, Association of American Geographers*, vol. 51 (1961) pp. 241–60.
22. Tuan, *Topophilia*, p. 11.
23. The pioneering work on personal space is Edward Hall, *The Hidden Dimension* (Garden City, New York: Anchor Books, 1969). The symbolic use of space can occur at larger scales than the personal. For example, in times of social stability, economic class may be spatially integrated residentially because class distinctions and differences are maintained by the social order in social rituals. But when the social rituals are not adhered to, class distinctions may then be expressed in the geographic separation of classes. Such issues are discussed by Lyn Lofland, *The World of Strangers* (New York: Basic Books, 1973); and Richard Sennett, *The Fall of Public Man* (New York: Knopf, 1977).
24. Mircea Eliade, *Australian Religions* (Ithaca: Cornell University Press, 1973) p. 42.
25. Tuan, *Topophilia*, p. 18.
26. Émile Durkheim and Marcel Mauss, *Primitive Classification* (University of Chicago Press, 1963).
27. O. Ortmann, 'Theories of Synaesthesia in the Light of a Case of Colour Hearing', *Human Biology*, vol. 5 (1933) pp. 155–211.
28. Ibid.
29. L. Riggs and T. Karwoski, 'Synaesthesia', *British Journal of Psychology*, vol. 25 (1934) pp. 29–41.
30. Ortmann, 'Theories of Synaesthesia in the Light of a Case of Colour Hearing'.
31. Arthur Symons, 'The Opium-Smoker', *Poems by Arthur Symons* (New York: John Lane, 1912) vol. I, p. 3.
32. William Shakespeare, *Twelfth Night*, I.i.1–6.
33. William Shakespeare, *A Midsummer Night's Dream*, v.i.194–6.
34. For a discussion of Marjory Pratt's depiction and the use of colour organs, see Susanne Langer, *Mind* (Baltimore: The Johns Hopkins Press, 1967) vol. I, pp. 183–5.
35. Riggs and Karwoski, 'Synaesthesia'.
36. Stephen Ullmann, *The Principles of Semantics* (Glasgow: Jackson, 1957) p. 280.
37. S. Newhall, 'Warmth and Coolness of Colours', *Psychological Record*, vol. 4 (1941) pp. 198–212.
38. Langer, *Mind*, p. 186. See also Deborah Sharpe, *The Psychology of Color Design* (Chicago: Nelson-Hall, 1974) for a review of the problems of

applying synaesthetic relationships.

39. The differences between the dynamic–affective and physiognomic ways of viewing were pointed out by Sylvia Honkavaare, 'The Psychology of Expression', *The British Journal of Psychology: Monograph Supplements*, 32 (1961) pp. 1–96. The implications of this distinction for geography are discussed by Yi-Fu Tuan, 'Nature Imitates Art: A Theme in Experiential Geography', in Donald Deskins *et al.* (eds), *Geographic Humanism, Analysis and Social Action: Proceedings of Symposia Celebrating a Half Century of Geography at Michigan* (Department of Geography, University of Michigan Publications No. 17, 1977) pp. 27–44.

40. Werner, *Comparative Psychology of Mental Development*, p. 73.

41. Heinz Werner and Bernard Kaplan, *Symbol Formation* (New York: Wiley, 1963) p. 208. These experiments seem to be about physiognomic seeing proper. See footnote 39.

42. Ibid., p. 209.

43. Ibid., p. 339.

44. Ibid., p. 341.

45. Ibid., pp. 340–1.

46. Ibid., pp. 348–9.

47. Ibid., p. 213.

48. Ibid., p. 350.

49. See symbol realism in ibid., pp. 35 and 251; and Stephen Ullmann, *Semantics* (New York: Barnes and Noble, 1964) pp. 80–115, where he discusses transparent and opaque words.

50. Robert Browning, *The Pied Piper of Hamlin*, VII, 11. 11–13.

51. Ullmann, *Semantics*, p. 83, referring to Milton's *Paradise Lost Book IX*, 1, 86, emphasis Ullmann's.

52. Ullmann, *Semantics*, p. 83, referring to Milton's *Paradise Lost Book X*, pp. 506–9, emphasis Ullmann's.

53. Alexander Pope, *An Essay on Criticisms*, in vol. I, *The Complete Poetical Works of Alexander Pope* (Boston: Houghton Mifflin, 1903).

54. For a discussion of this view see Ernst Cassirer, *Philosophy of Symbolic Forms* (New Haven: Yale University, 1955) vol. I; Werner and Kaplan, *Symbol Formation*, and Willbur Urban, *Language and Reality* (New York: Macmillan, 1939).

55. Werner and Kaplan, *Symbol Formation*, p. 104.

56. Ibid, p. 105.

57. Otto Jespersen, *Language: Its Nature, Development and Origin* (London: George Allen and Unwin, 1922) p. 405.

58. Herbert Clark, 'Space, Time, Semantics, and the Child', in Timothy Moore (ed.), *Cognitive Development and the Acquisition of Language* (New York: Academic Press, 1973) pp. 27–63.

59. Baldwin, *Theories of Child Development*, p. 237.

60. Jean Piaget, *The Language and Thought of the Child* (London: Routledge and Kegan Paul, 1959) 3rd edition, p. 13.
61. Ibid., p. 13.
62. Ibid., p. 16.
63. Werner, *Comparative Psychology of Mental Development*, p. 256, referring to Jean Piaget, *The Child's Conception of the World*, pp. 61–87.
64. Clark, 'Space, Time, Semantics, and the Child'.
65. Ronald Longacker, *Language and its Structure* (New York: Harcourt Brace and World, 1968) p. 18.
66. Werner, *Comparative Psychology of Mental Development*, p. 258; and Cassirer, *Philosophy of Symbolic Forms*, pp. 193–4; and Franz Boas (ed.), *Handbook of American Indian Languages* (Washington: Government Printing Office, Smithsonian Institution, Bureau of American Ethnology, Bulletin 40, Part 1, 1911) pp. 43, 112–16, 244–82 and 446.
67. Cassirer, *Philosophy of Symbolic Forms*, p. 192, referring to Humboldt.
68. Ibid., p. 193; and Edward Sapir, 'Study in Phonetic Symbolism', *Journal of Experimental Psychology*, vol. 12 (1929) pp. 225–39.

Chapter 6: Myth and Magic

1. There are, of course, many facets to magic and myth. Both seem to share a common pattern of reasoning which appears to the non-believer or non-practitioner as fantastic and opposite to the scientific. It is this aspect of myth and magic, and *religion* as well, that we are interested in and which joins these two patterns of thought together. Also, it should be borne in mind that while the examples of mythical–magical thought are taken primarily from non-Western societies, the Western world also has its share of such beliefs and not all of such beliefs are part of religion.
2. Ernst Cassirer, *An Essay on Man* (New Haven: Yale University Press, 1944) p. 81.
3. Alfred Howitt, *The Native Tribes of South-East Australia* (London: Macmillan, 1904) pp. 396–7.
4. Extracted from Franz Boas, *Tsimshion Texts* (New Series: Publication of American Ethnological Society Leydon: 1912) vol. III, pp. 71–146. For a structuralist analysis of this myth see Claude Lévi-Strauss, 'The Story of Asdiwall', in Edmund Leach (ed.), *The Structural Study of Myth and Totemism* (Edinburgh: Tavistock, 1967) pp. 1–47.
5. This is from Lévi-Strauss, *The Raw and the Cooked* (New York: Harper and Row, 1969), and refers to the second version of the Borroro myth, pp. 49–50.
6. We are concerned here only with the aspects of Lévy-Bruhl's theory that are about the 'pre-logical' form of thought in myth and magic. Lévy-Bruhl views this characteristic in the context of mystical sympathy which

we take to mean, in our context, that symbols are confused with their referents. Such a confusion need not be a problem in logic but rather a confusion in categories. For a discussion of purely logical problems in myth and magic see David E. Cooper, 'Alternative Logic in "Primitive Thought"', *Man*, vol. 10, New Series (1975) pp. 238–56. There is much more to Lévy-Bruhl's philosophy than the confusion of symbol and referent. For a review and evaluation of his philosophy see, for example, Evans-Pritchard, *Theories of Primitive Religion* (Oxford: Clarendon Press, 1965);and Robin Horton and Ruth Finnegan (eds), *Modes of Thought* (London: Faber and Faber, 1973). Among the many who have been influenced by his views is Ernst Cassirer.

7. Ernst Cassirer, *Language and Myth* (New York: Harper and Brothers, 1946) p. 33.
8. C. R. Hallpike in, 'Is There a Primitive Mentality?', *Man*, vol. 11, New Series (1976) pp. 253–70, draws attention to the fact that observations and psychological experiments suggest that the mental development of the primitive does not go far beyond the equivalent of Piaget's concrete operational stage. The reasons for this are the simple technical levels of these societies and also the fact that in primitive societies there are no schools which remove people from nature and which thus force them to rely on a self-conscious use of symbols.
9. Lucien Lévy-Bruhl, *Les Carnets de Lucien Lévy-Bruhl* (Paris: Presses Universitaire de France, 1949).
10. Claude Lévi-Strauss, 'Overture to the Crue et le Cuit', in Robert Georges (ed.), *Studies in Mythology* (Illinois: Dorsey Press, 1968) pp. 210–11.
11. In Claude Lévi-Strauss, *The Savage Mind* (University of Chicago Press, 1966) p. 268, Lévi-Strauss discusses the differences between his view and Lévy-Bruhl's.
12. For criticism of Lévi-Strauss see Marvin Harris, *The Rise of Anthropological Theory* (New York: Thomas Crowell Company, 1968) *passim*; and G. Kirk, *Myth: Its Meaning and Functions in Ancient and Other Cultures* (London: Cambridge University Press, 1974) pp. 42–83.
13. Claude Lévi-Strauss, 'Overture to the Crue et le Cuit', pp. 212–13.
14. Ernst Cassirer, *Philosophy of Symbolic Forms* (New Haven: Yale University Press, 1955) vol. II pp. 75–6.
15. Ibid., p. 91.
16. Ibid., p. 91.
17. Ibid., p. 91.
18. We have used Mauss' first two categories and incorporated his third principle, antipathy, within them; M. Mauss, *A General Theory of Magic* (London: Routledge and Kegan Paul, 1972) especially Chapter 3, Section 3.
19. Ibid., p. 64.
20. Ibid., p. 65.

21. Ibid., p. 68. We take this definition to include both 'like produces like, *similia similibus evocantur*, and like acts upon like, and in particular, cures like, *similia similibus curantur*.

22. Ibid., p. 68. Sometimes these need only be imagined or seen in the mind's eye. Imagination in the sense of visualisation plays an important role in magic.

23. We have reviewed these differences in Chapter 3.

24. Mauss, *A General Theory of Magic*, pp. 65–6.

25. Ibid., p. 67.

26. Henri Frankfurt *et al*, *Before Philosophy* (Baltimore: Penguin Books, 1949) p. 51.

27. Leslie White, 'The World of the Keresan Pueblo Indians', in Stanley Diamond (ed.), *Primitive Views of the World* (New York: Columbia University Press, 1960) p. 87.

28. Cassirer, *Philosophy of Symbolic Forms*, pp. 84–5.

29. Ibid., p. 88.

30. Mircea Eliade, *The Myth of the Eternal Return* (Princeton University Press, 1954) p. 18.

31. Douglas Fraser, *Village Planning in the Primitive World* (New York: Braziller, 1968); and Paul Radin, *The Story of the American Indian* (New York: Boni and Liveright, 1927), especially his discussion of the Pawnee Indians.

32. Andreas Volwahsen, *Living Architecture: Indian* (London: Macdonald, 1969) p. 44.

33. Joseph Needham, *Science and Civilization in China* (Cambridge University Press, 1962) vol. II, p. 359, quoting from Chatley, 'Feng Shui', *Encyclopedia Science*, p. 175.

34. Paul Wheately, *Pivot of the Four Quarters* (Chicago: Aldine, 1971) p. 419.

35. Chuen-Yan Lai, 'A *Feng-Shui* Model as a Location Index', *Annals, Association of American Geographers*, vol. 64 (1974) p. 508.

36. Arthur Wright, 'Symbolism and Function', *Journal of Asian Studies*, vol. 24 (1965) p. 671.

37. Wheately, *Pivot of the Four Quarters*, p. 435.

38. Celtic hill-fort structures may have been a form of incipient urbanism. See C. Smith, *An Historical Geography of Western Europe Before 1800* (New York: Praeger, 1967) p. 86.

39. Numa Fustel de Coulonges, *The Ancient City* (Garden City, New York: Doubleday, 1956) pp. 193–205.

40. Plutarch, *The Lives of the Noble Grecians and Romans*, The Dreyden Translation in Great Books of the Western World (Chicago: Encyclopedia Britannica Inc., 1952) vol. 14, p. 19.

41. M. Foucault, *The Order of Things* (New York: Vintage Books, 1970) p. 17.

42. Ibid., p. 26, quoting Paracelsus, *Die 9 Bücher der Natura Rerum* (Works (ed.), Suhdorff, vol. IX, p. 393).

43. Theophany in the history of Western attitudes towards nature is ˙discussed in C. Glacken, *Traces on the Rhodian Shore* (Berkeley: University of California Press, 1967) pp. 209–12; 238–9; and 282.

44. F. Yates, *Giordana Bruno and the Hermetic Tradition* (New York: Vintage Books, 1969) p. 132, referring to Agrippa, *De Occulta Philosophia*.

45. Glacken, *Traces on the Rhodian Shore*, pp. 203–4. According to Glacken, the view was that 'God is revealed in the Scripture; His works are also visible in the world. The book of nature is contrasted with the Bible, the book of Revelation, the former, however, [was thought by most as] being of a lower order than the latter because God is revealed in His word but only partly so in His works because he is a transcendent God.' To men like Sibiude, however, the book of nature was all-sufficient, composed of creatures in which 'a creature is a letter written by the finger of God, and many creatures, like many letters, make up a book', and was superior to the Scriptures because 'it cannot be falsified, destroyed, or misinterpreted; it will not induce heresy, and heretics cannot misunderstand it' (p. 239).

46. M. Kline, *Mathematics and the Physical World* (New York: Thomas Y. Crowell, 1959) p. 29. The Pythagoreans knew of five regular solids. They associated earth with the cube (for stability), water with the icosahedron, fire with the tetrahedron (simplest and lightest), air with the octohedron, and to the dodecahedron they assigned a mystery; H. Redgrove, *Magic and Mysticism* (New York: University Books, 1971) p. 18. The Pythagoreans also believed ten was ideal because it is the sum of one, two, three, four, and so they looked for ten heavenly bodies to reflect it. With one earth, the five planets, the sun, moon, and stars, they needed one more, which they found in an invisible 'counter earth'.

47. K. Seligmann, *The Mirror of Magic* (New York: Pantheon Books, 1948) p. 344.

48. For instance, several passages in Genesis tell how Abraham, with the help of 318 slaves, rescued the captive, Lot. Since the Lord was on Abraham's side, why were so many men needed? 'To this Gematria answers that the sum of the name of Abraham's majordomus, Eliezer, . . . of Damascus is three hundred and eighteen and that Abraham [therefore] defeated the four kings . . . with the help of one man.' K. Seligmann, *The Mirror of Magic* (New York: Pantheon Books, 1948) p. 348.

49. The origin of this early form of pharmacy has been traced to China; C. LaWall, *Four Thousand Years of Pharmacy* (Philadelphia: J. B. Lippincott, 1927) p. 19.

50. A. Wootton, *Chronicles of Pharmacy* (London: Macmillan, 1910) vol. 1, p. 185.

51. J. Matter, 'The Curious Doctrine of Signatures', *Modern Pharmacy*, vol. 1 (1943) pp. 6–7.
52. Yates, *Giordana Bruno and the Hermetic Tradition*, Chapter 6.
53. S. Moholy-Nagy, *Matrix of Man* (New York: Praeger, 1968) pp. 66–7. Outside of the West, innumerable towns were planned for magical purposes.
54. The following descriptions, based on the translation of *The City of the Sun*, are found in H. Morley, *Ideal Commonwealths* (New York: Colonial Press, 1901) pp. 141–79.
55. The City of the Sun as a memory devise is discussed in Yates, *Giordana Bruno and the Hermetic Tradition*, pp. 377–8.

Chapter 7: The Societal Conception of Space

1. Elmon Service, *Origins of the State and Civilization* (New York: W. W. Norton, 1975).
2. See Georges Gurvitch, *The Social Frameworks of Knowledge* (New York: Harper and Row, 1971) p. 25.
3. Stanley Diamond, *In Search of the Primitive* (New Brunswick, NJ: Transaction Books, 1974) pp. 124–5; and the *O.E.D.*
4. Elmon Service, *The Hunters* (Englewood Cliffs, NJ: Prentice-Hall, 1966) pp. 7–8.
5. Ibid., pp. 85–7.
6. Marshall Sahlins, *Tribesmen* (Englewood Cliffs, NJ: Prentice-Hall, 1968) pp. 31 and 40.
7. Ibid., p. 43.
8. Diamond, *In Search of the Primitive*, p. 167.
9. Ibid., p. 145.
10. Paul Radin, *Primitive Man as Philosopher* (New York: Dover Publications, 1957) pp. 32, 34 and 38.
11. Lucien Lévy-Bruhl, *The 'Soul' of the Primitive* (Chicago: Henry Regnery, 1971) p. 185.
12. See C. R. Hallpike, 'Is There a Primitive Mentality?', *Man*, vol. 11, New Series (1976) pp. 253–70.
13. Radin, *Primitive Man as Philosopher*, p. 52.
14. Ibid., p. 43.
15. Robert Redfield, 'Thinker and Intellectual in Primitive Society', in Stanley Diamond (ed.), *Primitive Views of the World* (New York: Columbia University Press, 1960) p. 39, referring to the Dogon.
16. Radin, *Primitive Man as Philosopher*, p. 72 See also Franz Boas, *The Mind of Primitive Man* (New York: Macmillan, 1921) pp. 216–19.
17. Diamond, *In Search of the Primitive*, p. 138.
18. Ibid., p. 192.

19. T. Strehlow, *Aranda Traditions* (Carlton, Victoria: Melbourne University Press, 1947) pp. 30—1.
20. Frank Speck, 'Penobscot Tales and Religious Beliefs', *The Journal of American Folk-Lore*, vol. 48 (1935) pp. 5—7.
21. See H. Alexander, *The World's Rim* (Lincoln, Nebraska: University of Nebraska Press, 1953) p. 89 for a discussion of Pawnee myth, and see Leslie White, 'The World of the Keresan Pùeblo Indians', in Stanley Diamond (ed.), *Primitive Views of the World* (New York: Columbia University Press, 1960) pp. 84—5.
22. Fr. Van Wing, *Etudes Bakongo*, douxième édition (Desclée de Brouwer, 1959) pp. 93—4.
23. Certainly plans had to be made about when to camp and so on, but these were concrete decisions about present activities and not speculations about the unfamiliar or novel. Their lack of speculation about these matters makes their conception of a future time quite different from ours.
24. For a discussion of these hypotheses see Service, *Origins of the State and Civilization*.
25. For all intents and purposes the city and the state are, historically, but two sides of the same coin.
26. Frederick Engels, *The Origin of the Family, Private Property, and the State* (New York: International Publishers, 1972) p. 229.
27. For those who agree and those who disagree on this point, see Laurence Krader, *Formation of the State* (Englewood Cliffs, NJ: Prentice-Hall, 1968) especially pp. 1—27.
28. The term social definition of territory was used in Edward Soja's *The Political Organization of Space* (Washington, DC: Association of American Geographers, Commission on College Geography Resource Paper, No. 8, 1971) p. 13. The idea is often attributed to Henry Maine, *Ancient Law* (New York: Henry Holt, 1888), but one can find references to it in Rousseau.
29. Robert Redfield, *The Primitive World and its Transformations* (New York: Cornell University Press, 1962) pp. 33—4.
30. Krader, *Formation of the State*, pp. 54—5, and Moret, *Le Nile el la Civilization Egyptienne* (Paris: Renaissance du lirre, 1926) pp. 47—67.
31. Engels, *The Origin of the Family, Private Property, and the State*, p. 176.
32. Ibid.
33. Ibid., p. 179. A more recent interpretation of this is found in Eric Voegelin, *Order and History*, vol. 2 (Baton Rouge: Louisiana State University Press, 1957) pp. 113—14.
34. Henri Pirenne, *Economic and Social History of Medieval Europe* (New York: Harcourt Brace and World, 1937) p. 51.
35. See David Harvey, *Social Justice and the City* (Baltimore: Johns Hopkins University Press, 1975) *passim*, for a discussion of absolute space in capitalism.

36. Plato, *Republic*, Jowett translation, Book VIII, 540.
37. Ibid., Book IV, 423.
38. Ibid., Book II, 414.
39. These and other views are reviewed in *The Challenge of Local Governmental Reorganization: Substage Regionalism and the Federal System* (Washington, DC: Advisory Commission on Intergovernmental Relations, 1974) vol. III, pp. 3–27.
40. These devices are defined and discussed in Robert Dixon, *Democratic Representation: Reapportionment in Law and Politics* (New York: Oxford University Press, 1968).
41. As Stephen Gale, 'A Resolution of the Regionalization Problem and its Implications for Political Geography and Social Justice', *Geografiska Annaler*, vol. 58 (1976) pp. 1–16, pointed out, areal representation has its drawbacks and it could be replaced in some cases by other forms of representation.
42. The example is Yi-Fu Tuan's. For a recent discussion of density and crowding, see Yi-Fu Tuan *Space and Place* (Minneapolis: University of Minnesota Press, 1977) pp. 51–66.
43. See M. Van Hulten, 'Plan and Reality in the Ijsselmeerpolders', *Tydschrift voor Economische en Sociale Geographie* LX (1969) pp. 67–76; and Jehoshua Ben-Arieh, 'The Lakhish Settlement Project: Planning and Reality', *Tydschrift voor Economische en Sociale Geographie* LXI (1970) pp. 334–47.
44. Edward Hall, *The Hidden Dimension* (New York: Doubleday, 1969); and Robert Sommers, *Personal Space: The Behavioral Bases of Design* (Englewood Cliffs, NJ: Prentice-Hall, 1969) discusses examples of rigid spatial design which are not functional. Discussions of disorderly physical environments which contain orderly and beneficial social environments are found in William Whyte, *Street Corner Society* (University of Chicago Press, 1943); and Elizabeth Bott, *Family and Social Network* (New York: Free Press, 1971).
45. Yi-Fu Tuan, 'Sacred Space: Explorations of an Idea', in Karl Butzer (ed.), *Dimensions of Human Geography* (University of Chicago, Department of Geography, Research Paper 186, 1978) pp. 84–99.
46. Gunnar Olsson, *Birds in Egg* (Ann Arbor: Michigan Geographical Publications No. 15, Department of Geography, University of Michigan, 1975) p. 466, discusses the dilemmas of planning.

Chapter 8: Geography and Social Science

1. John McMurtry, *The Structure of Marx's World-View* (Princeton University Press, 1978).
2. H. Harré and P. Secord, *The Explanation of Social Behavior* (Totowa,

New Jersey: Littlefield, Adams and Co., 1973) have discussed the role of ordinary language as a social science model.

3. See Derek Gregory, *Ideology, Science and Human Geography* (London: Hutchinson, 1978) for the dialectical relationships among viewpoints.

4. J. Entrikin, 'Contemporary Humanism in Geography', *Annals of the Association of American Geographers*, vol. 66 (1976) pp. 615–32.

Author Index

Subject Index